THE GREAT BLUE HERON

ROBERT W. BUTLER

THE GREAT BLUE HERON

A NATURAL HISTORY AND ECOLOGY OF A SEASHORE SENTINEL

UBCPress / Vancouver

Printed in Canada on acid-free paper ∞

ISBN 0-7748-0635-4 (hardcover)
ISBN 0-7748-0634-6 (paperback)

Canadian Cataloguing in Publication Data

Butler, Robert William.
The great blue heron

Includes bibliographical references and index.
ISBN 0-7748-0635-4 (bound); ISBN 0-7748-0634-6 (pbk.)

1. Great blue heron – British Columbia. I. Title.
QL696.C52B87 1997 598.3'4 C97-910694-X

This book was published with the help of a grant from Environment Canada.

UBC Press gratefully acknowledges the ongoing support to its publishing program from the Canada Council for the Arts, the British Columbia Arts Council, and the Department of Canadian Heritage of the Government of Canada.

UBC Press
University of British Columbia
6344 Memorial Road
Vancouver, BC V6T 1Z2
(604) 822-5959
Fax: 1-800-668-0821
E-mail: orders@ubcpress.ubc.ca
http://www.ubcpress.ubc.ca

For Sharon

CONTENTS

ILLUSTRATIONS

Figures

Photographs

Colour plates

FOREWORD

THERE ARE SEVERAL REASONS to welcome the work outlined in this timely book by Rob Butler. The two most important are, first, the spotlighting of the need to recognize races and subspecies in nature and, second, the emphasis on the need for humans to begin showing individual and group responsibility for protecting rapidly disappearing ecosystems.

Science has made great strides in genetic and molecular research in recent years. It is now becoming more and more evident that within any given species certain populations have evolved through thousands of years to fit particular ecological niches. For example, the white ash (*Fraxinus americana*) grows in a small area of northern Ontario. This boreal population is made up of naïve country bumpkins that expect warm weather to mean that spring has come. In this area, it is true. Spring comes suddenly, and buds burst open with an urgency appropriate to the short northern summer. However, their southern cousins are used to warm fronts bringing higher temperatures from time to time throughout the winter, and they are not fooled into bursting their buds until the spring really comes in May. If a northern tree is transplanted to the south, it will make the mistake of budding too soon and then get hit by frost. Thus, if a population is wiped out in one ecosystem, it may not necessarily be replaceable by the introduction of the same species from another ecosystem.

The more we understand the complexities of nature, the more we see endless complexities unfolding ahead of our understanding. Now that we know this from cell biology, we need to redirect our attention to whole animals and learn the intricacies not only of species but also of subspecies and races. If we are to preserve and protect biodiversity, we must know the names of all our neighbours of other species, and we must learn about their habits, needs, and idiosyncrasies. This is why Rob Butler's book is so important at this time. The great blue heron is found all over North America, and its close cousin, the grey heron, is found all over Europe, but coastal British Columbia has a particular subspecies. The populations of these birds are at risk due to human activities.

That point brings me to the second reason for the timeliness of this book – the need for stewardship. At the end of the twentieth century, virtually all human activities must be modified to reduce their impact on nature. Individuals, governments, and businesses need to change behaviour to ensure that nature is not degraded further and is even restored from the destruction of the past. This restoration is quite possible, but a philosophical shift will be required. E.F. Schumacher, who wrote *Small Is Beautiful*, has said that 'The real problems facing this planet are not economic or technical, they are philosophical.' A striking example is the growth of the human population: it is not an economic or technical problem but a philosophical one. Where there is a will, there is a way; where there is no will, there is no way.

If we protect the future of the herons, we protect ourselves, enhancing not only the life of future generations but also our own quality of life at present. This protection will cost money, but the best things in life are not free anymore. Clean air, clean water, and healthy wildlife populations used to be free; now we must pay for them with higher taxes or higher prices. We can pay now or later, but if we pay later, it will cost a whole lot more. Can anyone doubt that it is worth it?

In the economic realities of the 1990s, we can no longer expect large government budgets to bail us out of our ecological messes. That is where stewardship comes in. Individuals as well as businesses and volunteer groups can do an enormous amount with very little money. But hearts must be in the right place. Rob Butler says:

> Few other animals better symbolize a vision of conservation for the Strait of Georgia ecosystem than the great blue heron. It lives year-round on the shores of the strait, wades on its beaches and in its streams, rivers, and marshes, hunts in grasslands and from kelp forests, nests in old-growth rain forests, and penetrates the urban landscape. As sentinels, the heron's eggs provide a means to monitor contaminants in the rivers and ocean, and its reproductive success might just provide clues to the abundance of fish in inshore waters. Conserving

> the heron and its environment would go a long way toward ensuring the conservation of much of the quality of life in the Strait of Georgia and Puget Sound.

How true! In this book, he has given us the philosophical base that can make a great difference, not only for herons but for all of us.

Robert Bateman
Saltspring Island

PREFACE

> The Fraser's experience with man, while short, is filled with adventure, toil, treasure and war. Yet, unlike other great rivers, it has produced no songs, no myths, not even a special type of riverman.
>
> – Bruce Hutchison, *The Fraser*

So wrote Bruce Hutchison forty-five years ago, and how mistaken he was! For centuries, the lives and culture of the First Nations were inextricably tied to the Fraser River and its salmon-spawning runs. Before Europeans began to meddle in their lives, the First Nations along the Fraser River scheduled their year around the returning salmon. On this rich, reliable resource they built a complex culture, and their myths, songs, and art reflect this close bond to the river. Even the name one group gives to itself – the Stō:lo – simply means 'the river.'

When salmon were not in the river, the Stō:lo dug clams and snared crabs along coastal beaches, harvested eulachons in the rivers, and gathered herring eggs in coastal bays. In summer, they dug flower bulbs from forest clearings and picked berries along mountain slopes. Wherever they went, the Stō:lo were not far from the great blue heron.

In recent years, interest in restoring and maintaining the former biological wealth of the Fraser River drainage basin has become a prominent part of Canada's national environmental agenda. The significance of the Fraser River to fish was well established in the 1950s, and recently the international significance to birds has been recognized (Butler and Campbell 1987; Leach 1972; Butler and Vermeer 1989; Vermeer and Butler 1994). Today, the view that responsible resource use and conservation are essential ingredients of healthy resource-based economies is well entrenched. The Fraser River Action

Plan of Environment Canada and the Department of Fisheries and Oceans was a six-year venture launched in 1991 to restore and protect the river's ecosystems. This publication is one of the results of that program.

Much has been learned about the Fraser River, but we may never comprehend the immense ecological importance of the river to the life of the north Pacific. It is humbling to realize that the lives of grebes from the Canadian prairies, geese from Siberia, salmon from the north Pacific, sandpipers from Panama, and whales in the Strait of Georgia are tied to the same ecological processes that fuel the lives of herons on the Fraser River delta. Perhaps no species better symbolizes the Fraser River environment than the great blue heron, and ensuring its survival will go a long way toward preserving the life of this globally important delta and the Strait of Georgia.

The great blue heron is one of the most distinctive of North American birds. But for a species that has played such an important role in the lives of North Americans, much is unknown about its biology. Most thoroughly studied are aspects of its breeding biology and courtship behaviour. In recent years, much has been learned about its ecology on the British Columbia coast. My aim in writing this book is to bring together aspects of the biology and ecology of the great blue heron throughout its lifetime. Most of the information derives from studies by me and other students of herons on the coast of British Columbia. Although there has been some bias in choosing this location, it is the only place where the great blue heron has been studied throughout the year.

ACKNOWLEDGMENTS

Many people have helped me in writing this book. The research was funded by the Canadian Wildlife Service and supported by Art Martell from the CWS office in Delta and Keith Marshall and Tony Keith in Ottawa. The British Columbia Ministry of Environment, Lands and Parks, and especially George Trachuk and Rik Simmons, permitted and helped to establish the field camp on Sidney Island. Without their support, the Sidney Island project would not have been conducted. Kees and Rebecca Vermeer provided warm hospitality at their home in Sidney. Terry Sullivan helped to establish the camp on Sidney Island and enthusiastically collected data in the field for several summers. André Breault, Ian Moul, Phil Whitehead, Martin Gebauer, Don Norman, and the Vancouver Aquarium naturalists allowed me generous access to their data on heron colonies in British Columbia, and Hernan Vargas provided data on breeding herons from the Galapagos Islands. Ken Langelier kindly performed postmortems on nesting and juvenile herons. Many students and teachers helped to collect data on fish on Roberts Bank, including Rod MacVicar, Ruth Foster, Shannon Bennett, Robin Gutsell, Remco Tikkemeijer, Holly and Myrica Butler, David Suzuki, Jim Mattson, and students from the Fisheries Ecology Class from Centennial School: Krista Colton, Catherine Dickie, Lisa Fairley, David Galitzky, Tracy Mantison, Lisa Newton, Dave Tiessen, Daniel Tisseur, and Brent Vivian. Robin Gutsell and G.E. John Smith assisted with some data analysis, and Shelagh Bucknell showed me how little I knew about word processing. Ian Moul and Philip Whitehead kindly allowed me to use data of hatching dates by herons in the Strait of Georgia, and Paul Ferry provided data on heron numbers in the Point Roberts colony. Marg Holmes provided information on the archeology of First Nations people. Randy Burke of Bluewater Adventures and the crew of *Island Roamer* provided me with opportunities to visit many parts of the BC coast inaccessible by road. Many people collected data on birds at Sidney Island over the years. They included Moira Lemon, Terry Sullivan, Chris and Tina Schmidt, Rhys Bullman, Horacio de la Cueva, Jan Fennigan, Michael Price, Connie Downes, Rick Toochin, Scott Butler, Karla Tremaine, Sharon Butler, Holly Butler, Myrica Butler, and Ian Moul. Their enthusiasm made fieldwork that much more enjoyable. I benefited from discussions with Bob Elner at CWS; Ron Ydenberg, Dov Lank, and Fred Cooke at Simon Fraser University; and Jamie Smith at the University of British Columbia. Carolyn O'Neill, Dave Morgan, and Kim Joslin provided funds and enabled publication of the book. Margaret North allowed me to use a digital conversion of a historical map of the Fraser delta, and Jason Komaromi and Kathleen Moore did the conversion. Dov Lank and Dick Cannings provided insightful comments on the entire manuscript. Jean Wilson, Randy Schmidt, Holly Keller-Brohman, and Peter Milroy at UBC Press provided a smooth transition of the manuscript through to publication. Finally, I reserve special thanks for Robert Bateman, who kindly found time to write the foreword.

THE GREAT BLUE HERON

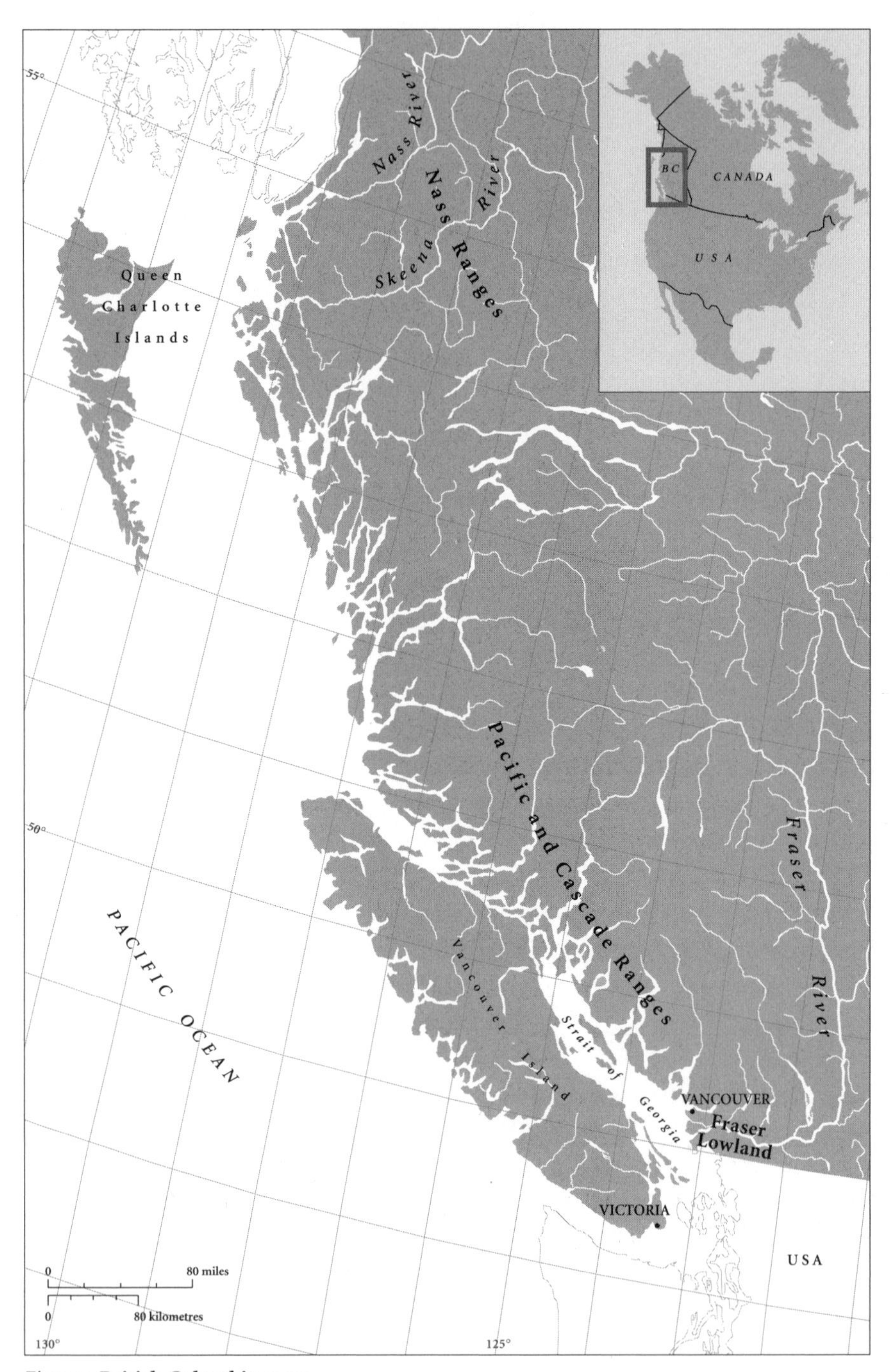

Figure 1 **British Columbia coast**

INTRODUCTION: THE SEASHORE SENTINEL

> The heron moves gently down the cove: step; pause; step. Watching it through binoculars and telescope, and occasionally writing in my notebook, I am deliberate and stealthy in my movements as it is, for I have been permitted onto the edge of the mystery, and I am amazed that this creature and I inhabit the same planet.
>
> – Michael Harwood, *Moments of Discovery*

WITHIN THE 800-KILOMETRE-LONG COASTLINE of British Columbia lie over 27,000 kilometres of fragmented shoreline. To the east of this shoreline stand the Nass, Pacific, and Cascade Ranges, and to the west lies the Pacific Ocean (see Figure 1). Bathed in Pacific rain and mist, this mysterious coastal strip is one of the most biologically diverse temperate regions on Earth (Harding and McCullum 1994). The superlatives that describe this coast are as impressive as the land itself – the world's largest intact temperate rain forests, the greatest salmon-spawning rivers, and the greatest number of bald eagles in winter, all set in some of the most spectacular shoreline on Earth. This temperate coastline is not the kind of place where one would expect to see herons. Riding on kelp beds, wading in coastal streams, stalking prey in grasslands, or nesting in the limbs of giant firs, the great blue heron seems anomalous along the wave-washed rocky shores of the province. Yet few species are as widespread and consistently present throughout the year as the heron. Herons evolved in tropical wetlands, far from the cooler, rocky shores of British Columbia. But over thousands of years, a unique subspecies of heron has evolved and become as much a part of the native fauna of British Columbia as the common raven, bald eagle, killer whale, and tufted puffin. And its year-round presence has made the great blue heron a sentinel of the changes we humans have brought to the coastal environment.

Heron populations are under threat in British Columbia. Urban sprawl has consumed quiet woodlots where the heron nests, its eggs and tissues contain contaminants from industrial, agricultural, and residential developments, and its nesting sites are increasingly disturbed by eagles and human activities. Nonetheless, herons continue to nest and feed in the busy urban environment – a testament to their adaptability. In an age of heightened environmental awareness, the heron has come to symbolize a healthy, tranquil environment. However, if we do not heed the changes we are imposing on our coastal environment, we stand to lose more than just the heron.

This book aims to provide a detailed account of the natural history of the great blue heron along the British Columbia coast and to demonstrate that conservation of the heron and its habitat ensures a diverse community of coastal species. The great blue heron is a worthy symbol of the conservation of coastal habitats. Much of this information is used to develop a conservation plan known as the Heron Stewardship Program, in which private and public landowners are encouraged to participate. The book closes with a look at the future and a hint that some good news is within our grasp.

CHAPTER 1
THE NORTHWEST COAST GREAT BLUE HERON

THE GREAT BLUE HERON is one of about sixty species of herons in the world. Other members of the heron family include egrets, bitterns, and night-herons. The two closest relatives are the cocoi heron, which inhabits South American swamps, and the grey heron, which lives along shorelines, rivers, and marshes in Europe and Asia and shows up periodically in the Caribbean islands and on the Atlantic coast of North America. These three species are very similar in their behaviour, morphology, and genetic makeup, leading some taxonomists to suggest they might constitute a single species. Since their ranges do not overlap, they are usually considered a 'superspecies' (Hancock and Elliott 1978; Payne 1979).

Taxonomists have identified between four and eight subspecies of great blue heron in North America, the Caribbean, Central America, and the Galapagos Islands. Hancock and Elliott (1978) recognized different subspecies occupying southeast North America, the Galapagos, middle North America, western United States, southern California, and Florida and the West Indies, plus two possible subspecies in Central America and the Caribbean. Payne (1979) recognized four subspecies. One subspecies sufficed for most of continental North America, with the exception of *fannini*, along the northwest coast of North America and the subject of much of this book, and *occidentalis*, inhabiting Florida and West Indies. The subspecies *cognata* on the Galapagos completed his list. The most distinctive subspecies is *occidentalis*, formerly given full species status and known colloquially as the great white heron. This all-white heron inhabits the mangrove swamps of southern Florida, the West Indies, and the Caribbean coast.

The decision to consider the great white heron as a subspecies of the widespread great blue heron marked the close of a long history in ornithology dating back over two centuries concerning the species of herons. In 1758, Linnaeus named the great blue heron as *Ardea herodias*, the 'heron heron,' after he read a description by George Edwards of an 'ash-colour'd heron from North America' (Harwood 1977). Later, John James Audubon described the great white heron in southern Florida as a new species. In 1858, Spencer F. Baird found an intermediate form of heron with the body of a great blue heron and the head of a great white heron, a form he described as a new species. A wave of discussion arose over whether this form was indeed a true species or a hybrid of the great blue and great white herons. For a while, it went by the name of Würdemann's heron. However, J.W. Velie discovered the evidence that settled the argument. In 1872 and again in 1875, he found nests in northern Florida, where only great blue herons bred, containing young great white herons and Würdemann's herons or southern great blue herons. This was all the evidence that taxonomist Robert Ridgway needed, and in 1880 he considered Würdemann's heron a hybrid of the great blue heron. Ridgway separated the great white heron from its smaller close relative, the great blue heron, at the subspecies level based on evidence of regular interbreeding and size differences. Not everyone agreed with this decision. Arthur Cleveland Bent (1926) wrote: 'I cannot understand how anyone who is familiar with the great white heron in life can have any doubt that it is a distinct species. It is a strictly maritime species, its habitat is decidedly restricted and its behaviour is quite different from the Ward [great blue] heron, with which it mingles in the Florida Keys and doubtless interbreeds' (99).

The subspecies of great blue heron on the Gulf of Mexico and on the Galapagos share many of the same features (see Table 1). They breed year-round, lay small clutches of eggs, forage and nest in similar habitats, and are sedentary. However, captive *occidentalis* herons (great white herons) are very aggressive toward great blue herons and humans and their pets, a feature apparently not shared by great blue herons (Bent 1926). The continental subspecies has a distinct breeding season, lays relatively large clutches, and forages in many wetland habitats, and northern populations are migratory. The Pacific Northwest subspecies of great blue heron shares features of all three subspecies. Like the two southern subspecies, it is sedentary and forages along seashores; like the continental subspecies, it has a distinct reproductive season and lays relatively large clutches. It received its subspecific name *fannini* in honour of John Fannin, a former director of the British Columbia Provincial Museum in Victoria. The type locality is the Queen Charlotte Islands.

Table 1

Features of four subspecies of great blue herons*

	A.h. fannini	*A.h. herodias*	*A.h. occidentalis*	*A.h. cognata*
Location	Pacific Northwest	Continental N. America	Gulf of Mexico	Galapagos
Seasonal dispersion	Sedentary	Northern populations migrate south	Sedentary	Sedentary
Degree of sociality	Mostly colonial	Mostly colonial	Mostly colonial	Mostly solitary pairs
Breeding season	March-August	January-August (later start in the north)	Year-round	Year-round
Clutch size	4	3-5	2	2-3
Diet	Fish, small mammals	Fish, amphibians	Fish	Fish, iguanas, lizards, and hatching sea turtles
Foraging habitat	Mostly seashore	Fresh and saltwater	Seashore	Mostly seashore

* Subspecies names follow Payne's (1979) nomenclature.

Morphology of the Heron

The great blue heron is the largest heron in North America, with a body mass of more than twice that of any other species of heron. It lays the largest volume egg (seventy-five grams) and has the smallest clutch mass as a percentage of female body mass (10.3 per cent). The sixty-day period of nestling dependence on adults is the longest among North American herons. As with many birds, the mass of clutches and eggs, length of nestling period, and maximum life span of herons are closely related to the size of females. Large-bodied species tend to lay relatively small clutch masses, have the longest incubation and nestling periods, and live the longest.

The sexes of great blue herons are indistinguishable by plumage colour. However, most males are larger than most females. John Kelsall and Keith Simpson (1979) examined the gonads of three live herons and eighteen dead herons found during the winter in British Columbia. For most measurements, males were about 5 to 15 per cent larger than females. Keith Simpson (1984) analyzed these data and showed that the length of the culmen (or bill) was a good feature for distinguishing the sexes. The conventional way of measuring the culmen is between the bill tip and the point where feathers emerge at the forehead. Great blue herons have a small region of featherless skin where the bill and forehead meet, and this is the point where culmen lengths are measured. Simpson showed that the culmen lengths of male herons ranged from 129 to 146 millimetres and of female herons from 112 to 131 millimetres. Since there was only a small amount of overlap, culmen lengths are a good way to distinguish between the sexes. In British Columbia, herons with culmens longer than 131 millimetres are most likely male, while herons with culmens shorter than 129 millimetres are most likely female. With practice, the differences in culmen lengths between individual birds become quite evident, especially when mated pairs stand side by side at the nest. Milstein, Prestt, and Bell (1970) used this feature in their pioneering study of the grey heron in England. Butler, Breault, and Sullivan (1990) devised a method to assign sex to live herons at a distance by comparing the lengths of culmens measured against a graticule scale imbedded in the eyepiece of a telescope.

In the field, differences in plumage provide a reliable method of distinguishing fledgling, juvenile, and yearling herons from adults. The most obvious features of the adults are the white crown, long body plumes, grey-blue wing coverts, and black 'epaulets.' At a distance, the white crown and black epaulets are visibly striking. I was able to distinguish these features from an aircraft flown at 140 kilometres per hour during low-level flights. During the first few weeks after leaving the nest, herons are referred to as fledglings. They can be distinguished from other age classes by remnants of down protruding from a slate-grey crown. Features shared with other age classes are pale blue eyelids, yellow irises, and legs and toes that are olive green on the top and leading edge and pale greenish yellow below. For the rest of the first year of life, young herons are referred to as juveniles. They sport grey crowns and chestnut-coloured edging to wing coverts, and they lack the long body plumes or shoulder patches of adults. Herons between twelve and twenty-four months of age are referred to as yearlings. They wear an intermediate plumage in which the crown stripe is small, body plumes are short, a few coverts are edged chestnut, and a small shoulder patch is present. Juveniles can be easily distinguished in the field from other age classes, but yearlings are easily mistaken for adults. The plumages of males and females are identical in appearance in all age classes.

Herons in British Columbia

About 18,000 years ago, North America and Asia were connected by a 1,000-kilometre-wide land bridge known as Beringia. British Columbia lay beneath more than one kilometre of ice and snow. About 15,000 years ago, the great ice sheet began to retreat from the coast, opening a corridor along the coast of British Columbia. Glacial soils carried by the meltwater rushed to the sea, creating the beginnings of river deltas. Into this new land ventured animals from the south and north. The mild climate and isolation hemmed by sea and mountains set the stage for the evolution of new species and subspecies. There are endemic species of sticklebacks in coastal lakes, unique salamanders in forests, and a coastal-dwelling crow (Cannings and Cannings 1996).

Among the early immigrants to the coast were Aboriginal people, who arrived on the newly formed Fraser River delta about 9,000 years ago. They left no written history, but we know something of their lives from the remains and artefacts left in middens or garbage pits near their dwellings. Among the remains of shellfish, mammals, fish, and birds found in middens on the Fraser River delta and along the southern Strait of Georgia are bones of the great blue heron (Hobson and Driver 1989). Some long bones were shaped into whistles and drinking straws. One of the most remarkable items is a pestle shaped into a heron from the Marpole site that dates between 500 BC and AD 500. Given that only a handful of carvings survived from these early times, it is remarkable that the most beautiful one is of a heron.

This is among the earliest carvings showing the northwest coast art style. However, the heron is surprisingly scarce in ritual ceremonies. This oversight seems unlikely given the long history of the heron in coastal habitats and the widespread animal mythology of the Aboriginal people. It seems improbable that such a majestic animal as the great blue heron could be ignored. Instead, I suspect that the heron has been there all along but has been misidentified as a 'crane' by anthropologists. Sandhill cranes were widespread when Europeans first explored the south coast of British Columbia, and their references to cranes were likely inclusive of any long-necked, long-legged wading bird. However, specific references to herons were also made by explorers. In 1792, the Spanish-led Sutil and Mexicana expedition set sail from Nootka

Sound to explore the Strait of Georgia. On board was Don Joseph Mozino Suarez de Figaro, who was to consider the less serious matters of the journey. In his 'Noticias de Nutka' are references to Native people: 'On gala days [they] throw over the hair many small white feathers which they pluck from Eagles and Herons' (Pearse 1968). Captain King on the Cook expedition (1776-9) also noted the depiction of herons by Native peoples: 'In their dances, they disguise themselves with wooden masks representing figures of aquatic birds that they imitate' (Pearse 1968). Archibald Menzies on the Vancouver expedition of 1792 reported seeing herons on 6 June on Strawberry Island in the Strait of Georgia (Pearse 1968), and John Hoskins on the same voyage referred to herons not being abundant on the Queen Charlotte Islands and in Clayoquot Sound, a situation that remains today (Pearse 1968).

Given that herons masqueraded as cranes, what role did 'cranes' have in Native culture? Among the most remote coastal people are the Haisla and Heiltsuk bands who inhabit the steep fjords and canals of the central coast in the vicinity of the town of Bella Bella. Their rich culture included depictions of mythical animals, some with clear connections to real animals. Their year was divided into *Bakoos* (the ordinary work season, corresponding with spring, summer, and autumn) and *Tseka* (the winter dance season, when the world of the spirits drew close to the Haisla and Heiltsuk). People transformed into animals, and mythical beings appeared from the woods. Khenko was one of those mythical animals, referred to as a supernatural crane by anthropologists untrained in ornithological taxonomy. The distinctive head plumes, black eyebrow, and dagger-shaped bill of the masks are clearly those of the heron rather than the crane. The role of Khenko in the dance ceremonies is unclear, but its image is clearly depicted on ceremonial masks, bracelets, and amulets, as well as in day-to-day items such as net floats and canoe bailers. It is not difficult to imagine how Aboriginal people seeking spirits in the rain-soaked forests of the Pacific Northwest coast would be stirred by the harsh croaking call and rush of air pumped by the outstretched wings of a heron startled at night from a deep forest.

For more recent immigrants, with cultures far removed from the Pacific coast environment, the relationship with the heron has been mixed. Herons have been vilified as pests at fish-rearing ponds yet have become a symbol of wetland conservation. The heron's eggs and tissues have been used as indicators of industrial contamination, and its feathers have been used as adornments by the millinery trade. The slaughter of herons and egrets for their feathers, and other waterbirds for food, at the turn of the century galvanized public opinion to create the first international treaty for bird conservation.

Based on fossil evidence, herons date as early as the Miocene epoch and probably earlier. The oldest record for the genus is a 14-million-year-old specimen unearthed in the Observation Quarry in Nebraska. The Miocene epoch was characterized by species radiation of mammals and flowering plants between 5 and 23 million years ago. Many fossilized specimens of great blue herons in the United States and the West Indies date to the Pleistocene epoch 1.8 million years ago, when humans first appear in the fossil record. On the British Columbia coast, heron bones have been dated to at least 1,500 years ago (Hobson and Driver 1989), but herons probably inhabited the coast long before then. The suite of species in the garbage pits or middens of Aboriginal people includes the great blue heron; three species of loons; five species of grebes; three species of cormorants; twenty-six species of ducks, geese, and swans; six species of gulls; three species of alcids; the northwestern crow; and the bald eagle. All these bird species are found on the shores of the Strait of Georgia and in the Fraser River delta today.

Distribution and Number of Herons

The present worldwide breeding range of the great blue heron stretches from western Alaska in the north to the Galapagos Islands in the south (see Figure 2). Coastal British Columbia populations are distinctive in that they do not migrate, unlike herons in the interior of the province and elsewhere in Canada. Some local movements to foraging areas occur, but there is no exodus of herons from the coast in winter. It is unlikely that many herons from the interior of the province spend the winter on the coast. Most interior herons breed south of the Fraser River and the most direct corridor through the mountains to the coast. As a result, herons on the coast remain isolated year-round from populations that migrate. This isolation has promoted the local adaptation of coastal great blue herons, the most noticeable being a darker plumage that has prompted a distinctive subspecies classification, *Ardea herodias fannini.*

Estimating the number of herons on the British Columbia coast is a difficult task because of the terrain. The Canadian Wildlife Service inventory of all historical records of colonies in British Columbia between 1921 and 1985 listed eighty-four sites on the coast (Forbes et al. 1985a). More recent censuses showed that there were five large nesting colonies with over 100 pairs in

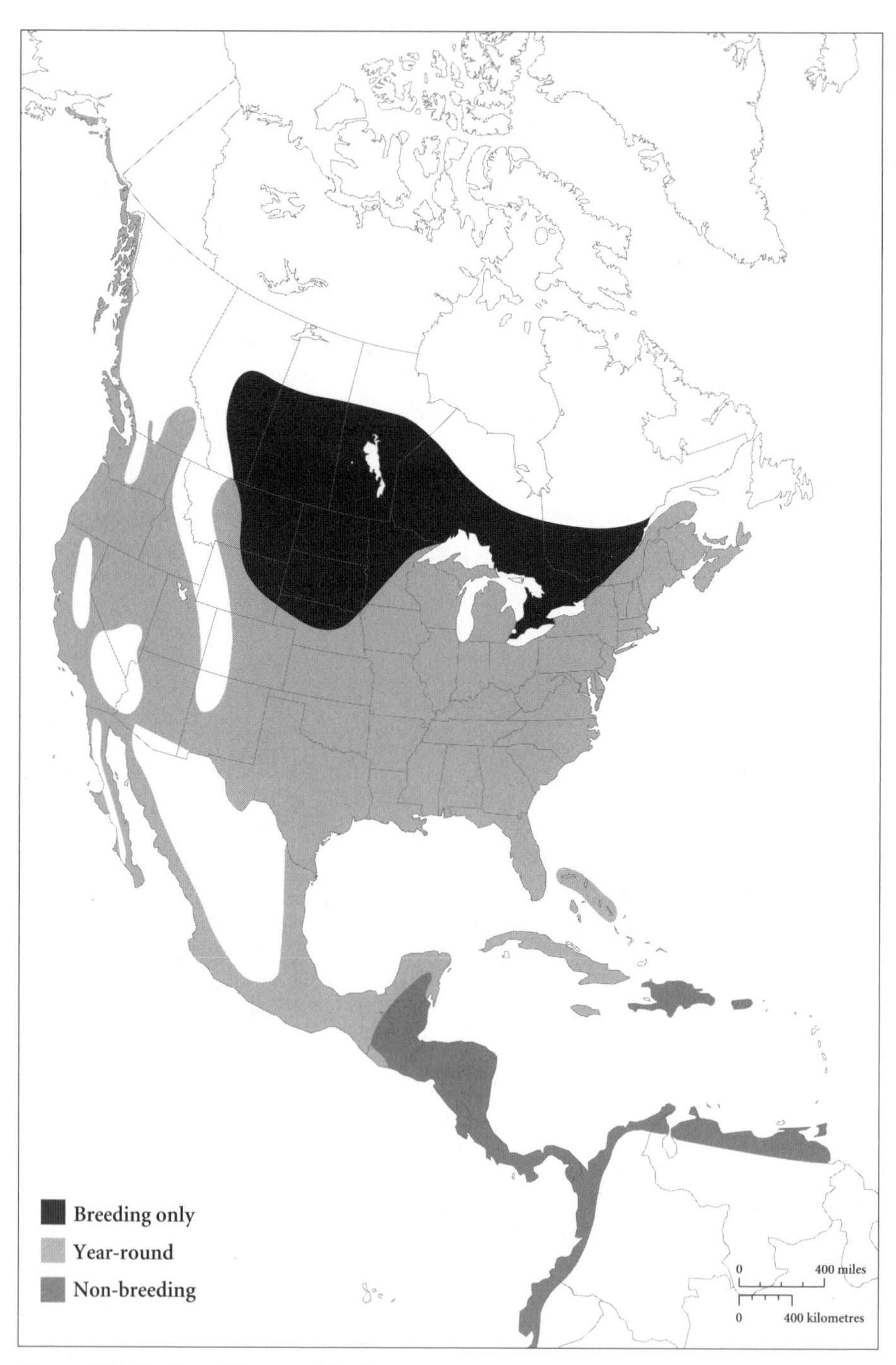

Figure 2 **Distribution of the great blue heron**

British Columbia between 1988 and 1992 (see Table 2): Holden Lake on Vancouver Island, Sidney Island in the Gulf Islands, Pacific Spirit Park in Vancouver, Point Roberts in the Fraser River delta, and the Canadian Forces Base near Chilliwack in the Fraser Valley. The Point Roberts herons, although nesting in Washington State, forage exclusively on the Fraser River delta. About 970 nesting pairs are confined to colonies with 100 or more nesting pairs in the Strait of Georgia and 1,106 in colonies with fewer than 100 pairs for a total of about 2,100 pairs (see Table 2). As many as 117 sites are known to have been used in Puget Sound between 1978 and 1994 (Norman 1995), and all large nesting colonies have likely been found given their conspicuous presence. The total number of herons in the five large colonies in Washington is 768 pairs (see Table 3). About half the herons around the Strait of Georgia and Puget Sound nest in large colonies (Butler et al. 1995; Eissinger 1996).

Table 2

Estimates of the number of breeding pairs of great blue herons (subspecies *fannini*) in BC colonies by colony size

Location	≥100 pairs	<100 pairs	Reference
Vancouver Island	265	431	Butler et al. 1995
Fraser River delta	553*	40	Butler et al. 1995
Fraser River valley	152	635	Gebauer 1993
Total	970	1,106	

* Includes colony of 387 pairs on Point Roberts, WA, that forages on the Fraser River delta.

Table 3

Estimates of the number of great blue herons (subspecies *fannini*) breeding in colonies with >100 pairs in Washington*

Location	Number of pairs
Thurston County, Tenino Road	100
Whatcom County, Birch Bay	200
Skagit County, Samish Island	210
Skagit County, March Point	150
King County, Maury Island	108
Total	768

* Data from Norman (1995).

About 2,400 pairs of herons might have nested around Puget Sound in the early 1990s (Eissinger 1996). Subtract 400 pairs on Point Roberts, Washington, that I included in the Fraser River delta (see Table 2) and that leaves about 2,000 pairs in Puget Sound. Therefore, the total minimum breeding heron population for the Strait of Georgia and Puget Sound is estimated to be about 4,100 pairs. Pairs nest sporadically along the west coast of Vancouver Island and along the coast as far north as Prince William Sound, Alaska (Campbell et al. 1990). I have never seen or heard of any large flocks in this region during the breeding season. A best guess might be a pair of herons on every 100 kilometres of the 27,000-kilometre-long coastline of British Columbia, for a total of about 300 pairs. A similar number might occupy the Alaskan coastline, for a total of about 600 pairs outside the Strait of Georgia and Puget Sound. Therefore, the total number of nesting herons of the subspecies *fannini* in Washington, British Columbia, and Alaska is probably fewer than 5,000 pairs (10,000 individuals). From this analysis, a picture emerges of the importance of the southern Strait of Georgia, the lower Fraser River valley, and northern Puget Sound to the survival of this subspecies of heron – over 80 per cent of all individuals occupy this region.

There have been nine major studies of great blue herons on the coast of British Columbia (see Table 4). Most were conducted during the breeding season and dealt with topics such as factors affecting breeding success, growth of chicks, contamination levels, and food. My studies followed herons throughout the year and addressed the timing of breeding, foraging, survival, and breeding success. The length and breadth of these studies are testament to the importance placed on the great blue heron as a species in British Columbia.

Table 4

Major studies of great blue herons in British Columbia

Location	Number of years	Main subjects	Researchers
Fraser River delta and Sidney Island	6	Breeding, foraging, habitat selection, survival	Butler (1989, 1991, 1995)
Pender Harbour	4	Breeding biology	Kelsall and Simpson (1979); Simpson (1984); Simpson et al. (1987)
Strait of Georgia	6	Reproductive success	Butler et al. (1995)
Crofton and Sidney Island	2	Breeding behaviour	Moul (1990)
Fraser River delta	2	Foraging of age classes	Gutsell (1995)
Fraser River delta	2	Growth of chicks	Bennett (1993); Bennett et al. (1995)
Strait of Georgia	4	Contamination	Elliott et al. (1989, 1996); Whitehead (1989)
Strait of Georgia	7	Reproductive success	Forbes et al. (1985b)
Fraser River delta	2	Foraging behaviour, colonial breeding	Krebs (1974)

A Word about Statistics

I wrote this book with the naturalist, ecologist, and conservationist in mind. For the naturalist, I included the natural history of the heron throughout the year. Statistics are a necessary tool of research and will be welcomed by some readers, but I have kept the statistics in this book to a minimum. Readers interested in more technical information can usually find it in the publications cited in the text. However, in some cases where new information is provided, statistical information is included. Readers not familiar with statistical symbols can find them in any introductory statistics text.

CHAPTER 2

THE COASTAL REALM OF THE GREAT BLUE HERON

THE COASTAL GREAT BLUE HERON resides along the shores of the temperate rain forests of western North America. Protected from the pounding surf of the open Pacific Ocean by Vancouver Island and sheltered from icy winter weather by the Coast and Cascade Ranges, this region provides the calm water, wide beaches with abundant fish, and mild weather to support herons throughout the year. The centre of the coastal range is the southern Strait of Georgia and northern Puget Sound (see Figure 3). Within the Strait of Georgia, the great blue heron shares habitats used by nearly one-quarter of the species of birds in Canada and one-third of the species found in British Columbia. In stark contrast to the sedentary heron, many of these species will grace the beaches and lands of other nations in the Western hemisphere for part of the year. Herons on the beaches, intertidal marshes, and farmlands of the Fraser River delta witness in spring and fall a tremendous number of waterfowl and shorebirds (see Figure 4) migrating between breeding haunts in Alaska, Siberia, and northern Canada and winter quarters as far away as South America. The heron forages in winter along the same rocky shores that provide food for one of the world's most important sites for seaducks, the sheltered waters of the Gulf Islands (see Figure 4). For two months in late spring and early summer, the Strait of Georgia is empty of the large flocks of waterfowl and shorebirds, and the predominant species that share the realm of the heron are seabirds from nearby nesting islands in the Strait of Georgia (see Figure 4).

The climate in the Strait of Georgia appeals not only to birds – one out of every fifteen Canadians lives in the region. And our lifestyle is often in conflict with herons and their ecosystems. The international destinations of birds

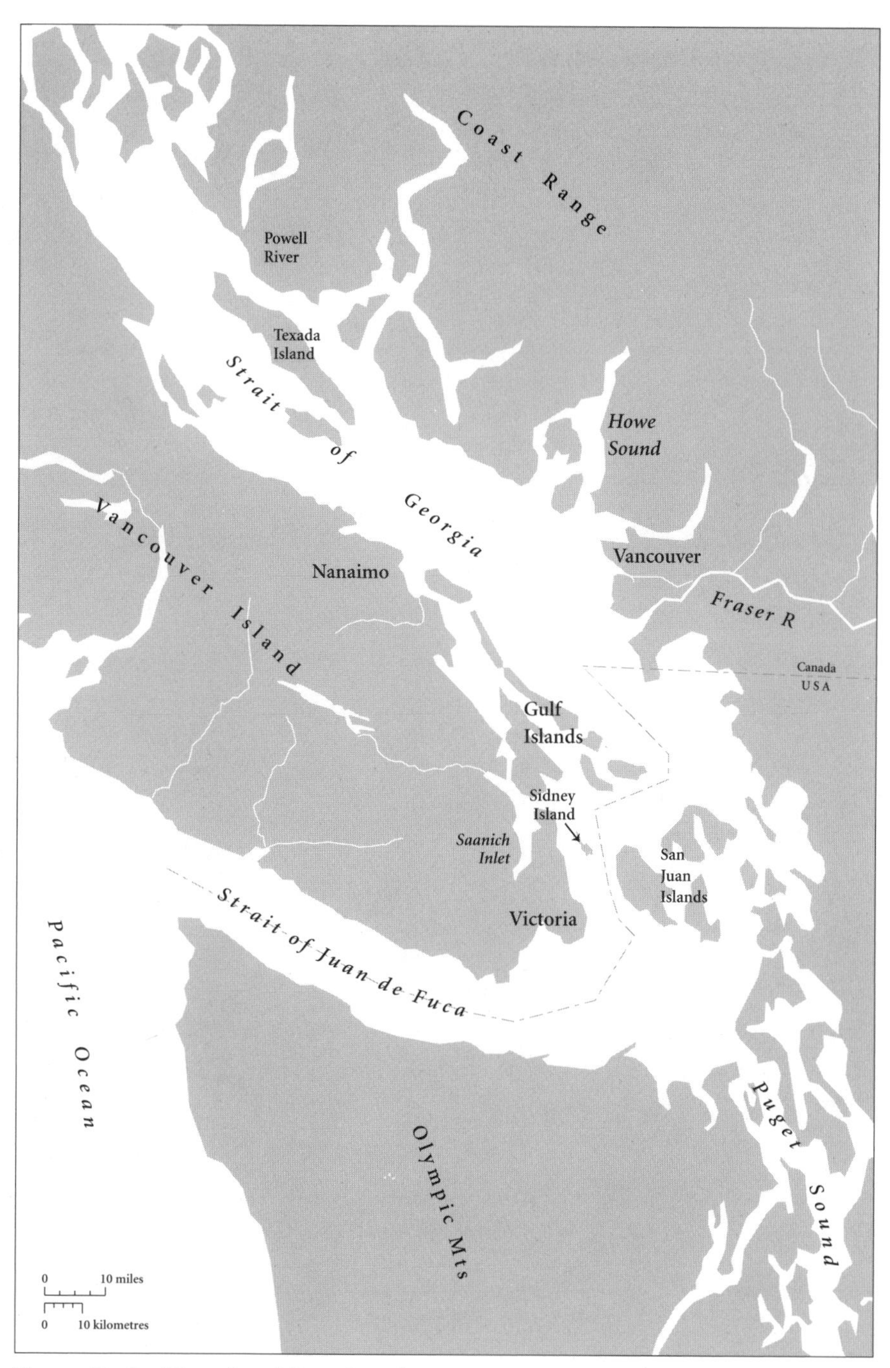

Figure 3 **Strait of Georgia and Puget Sound**

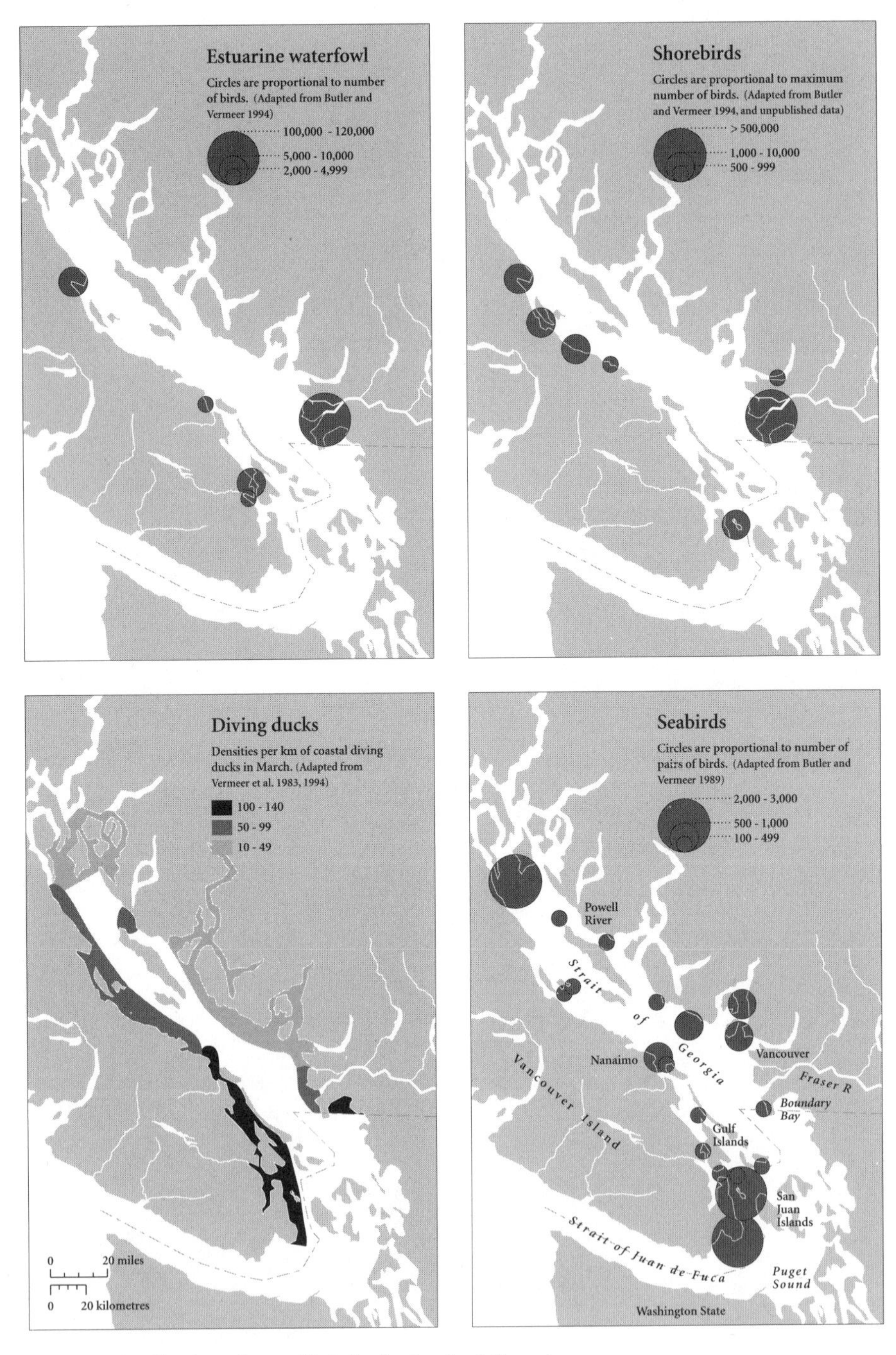

Figure 4 **Distribution of waterbirds in the Strait of Georgia**

that pass along the shores of the Strait of Georgia each year are a clear reminder that what we do to our ecosystems will be felt across the Western hemisphere. The immense responsibility to maintain healthy ecosystems underscores the importance of understanding the ecology of the Strait of Georgia. One of the original aims of the heron studies by the Canadian Wildlife Service was far less lofty than this but no less important.

A by-product of some kraft pulp mill operations is the production of small quantities of chlorine-based chemicals known as dioxins. There are many different forms of dioxins, and some are more toxic to wildlife than others. In the late 1980s, some toxic forms of dioxin contaminants were discovered in heron eggs collected from nests around the Strait of Georgia by my colleague Phil Whitehead. He was concerned that widespread contamination in the Strait of Georgia ecosystem might have occurred, but little was known about herons or their food in the region. Equally puzzling was how dioxins found their way into heron eggs. Some contaminants found in heron eggs are known to enter food webs by binding to suspended detritus particles. These contaminated particles are then consumed by invertebrates such as clams and shrimp, which are eaten by fish and birds, including the heron. Minuscule amounts of contaminants are readily concentrated in invertebrates that consume immense quantities of detritus. With each subsequent consumer, the concentration grows until eventually large quantities reside in heron tissues. Many contaminants are stored in fatty tissues. When female herons mobilize fat to produce eggs, the contaminants travel from tissues into the eggs. This route of contamination was most likely the means by which dioxins got into heron eggs around the Strait of Georgia.

However, there was a problem with this hypothesis – contaminant levels in prey recovered from heron nests with young chicks were lower than expected given the levels found in eggs from the same nests. This discrepancy might be explained by the time of collection and the source of prey fed to chicks and used to make eggs – prey used to make eggs would have been eaten in March and April, whereas prey fed to chicks would have been caught in May. If contaminants were accumulating in heron eggs, they were likely also concentrating in other animals that occur alongside the heron. Questions such as these had no answers and prompted more questions. Since contaminants such as dioxins travel through ecosystems by way of food, a sketch of who eats whom would provide a roadmap of the contamination of an ecosystem. Ecologists refer to these 'roadmaps' as food webs.

On the Fraser River delta, the great blue heron forages on farmlands, marshes, and mudflats. During migration and in winter, these habitats are also used by hundreds of thousands of waterfowl and shorebirds, such as the western sandpipers shown here. *(Photographs by Rob Butler)*

The Heron's Food Web

The lives of herons on the British Columbia coast are tied to a diverse marine ecosystem. Much of their food is derived from eelgrass and estuarine ecosystems fuelled by energy released during the disintegration of their leaves (Simenstad 1983). During growth, plants harness energy from the sun and store it in the form of carbohydrates in their leaves, stems, roots, and rhizomes. Some of this food energy enters the avian community directly through grazing by waterfowl, but most enters it via a complex food web that links bacteria, plankton, clams, crabs, and fish with mammals and birds, including the heron. Eelgrass provides a secondary function by lending support to a community of epiphytic plants and animals that gets its food energy from the sun or plankton in the water. At the same time, the suspended community provides food and shelter from predators for young fish and acts as a sink for nutrients suspended in the seawater. Two species of eelgrass live on the beaches of the Strait of Georgia; *Zostera marina*, a native species with wide fronds reaching nearly a metre in length, inhabits the lower and subtidal beach, while an introduced species, *Zostera japonica*, with narrow fronds seldom growing more than twenty centimetres in length, inhabits the upper intertidal beach. Both species have become widespread, with exceptionally fine examples in Boundary Bay and south Roberts Bank on the Fraser River delta, on Sidney Island in the Gulf Islands, and in Comox Harbour on Vancouver Island.

Eelgrass prefers clear seawater and temperatures above about 6°C for growth. In March, the low tides begin to uncover the beaches in the Strait of Georgia during the day. The increased day length results in a warming of the beach and the water in the eelgrass meadows. Consequently, growth of eelgrass fronds begins in March, and by May the meadows are a luxuriant green colour. Concurrent with the growth of fronds is the appearance of plants known as epiphytes, which use eelgrass fronds to support themselves near the water surface. The epiphytic community includes algae and a large variety of invertebrates.

The large number of fish attracted to eelgrass meadows is the mainstay in the diet of the heron, but these fish are of little commercial value, and their natural history is not well known. Gordon and Levings (1984) inventoried the fish in eelgrass meadows on the Fraser River delta over a period of a year. On Roberts Bank, they caught fifty-two species of fish in eelgrass meadows using a beach seine net pulled through the shallows. The predominant species caught were Pacific herring, staghorn sculpin, sandlance, shiner perch, and tube-snout.

The great blue heron is part of a complex food web that begins with estuarine marsh plants that harness the sun's energy. So important are the marshes that without them, the abundance of fish, birds, mammals, and invertebrates that occurs in estuaries would be greatly diminished. *(Photograph by Rob Butler)*

Eelgrass meadows provide herons with important fishing habitat. Most of the large heron colonies are located near eelgrass meadows. *(Photograph by Rob Butler)*

On Sidney Island, my beach seine sampling from April to July 1987 and 1988 caught twenty-three species. The predominant species were gunnels, shiner perch, staghorn sculpin, pipefish, and sticklebacks. Many of these species are important prey for herons.

Each spring, huge schools of shiner perch leave the deep waters of the Strait of Georgia to reproduce in the eelgrass shallows. On summer evenings when the tide is high and the air is calm, feeding shiner perch ripple the surface of the lagoon. Female perch are viviparous, and when they arrive in the eelgrass meadows in late April and May, their bellies are noticeably distended. Female shiner perch begin to reproduce when they are about 60 millimetres long. By June, young shiner perch are plentiful in the eelgrass meadows on Sidney Island, and the seasonal bounty attracts not only herons but also flotillas of double-crested cormorants, small parties of hooded mergansers, and lone belted kingfishers. Even bald eagles get in on the bounty. They sweep across the lagoon to snatch perch from just below the water. Cormorants dive in unison, driving the perch into shallow waters where they can catch them. On some days, scores of herons line the shore awaiting the diving cormorants.

Gunnels are numerous in eelgrass meadows, and samples taken from April to July 1987 and 1988 indicated that the average size of fish increased from about sixty-one to seventy-four millimetres during this period. Whether this represented growth of individuals or larger fish entering the lagoon was not clear. The region just below the surface of eelgrass meadows is the most biologically rich region of the ecosystem since most of the life is concentrated there. For many fish, survival and fecundity are dependent on size: the larger a fish, the greater its chances of not being eaten by predators and the more offspring it can produce. For a gunnel, the region near the surface is the place to be for rapid growth because most of its food is concentrated in the eelgrass canopy. However, this region is where gunnels are most vulnerable to herons. Young-of-the-year gunnels occupy the eelgrass canopy (Hughes 1985), where the risk of falling prey to a heron is probably greater than in deeper water.

Eelgrass meadows in the Strait of Georgia are used as nursery areas for immense schools of Pacific herring in February and March. Tens of thousands of feasting gulls, diving ducks, loons, grebes, and sea lions have been seen in some bays with spawning herring. Throughout March and April, brant from Mexican winter quarters join the remnant populations that winter in the strait on migration to the Arctic. By late April, hundreds of thousands of shorebirds, principally western sandpipers, probe the mudflats in search of invertebrate food. A large proportion of the world's population of about 5 million western sandpipers passes across the Fraser River delta on migration.

In autumn and winter, herons can often be seen catching meadow voles in grasslands in the Fraser River delta. Comparatively more is known about the natural history of small mammals in grasslands on the Fraser River delta than about fish in eelgrass meadows and estuarine marshes. Pioneering studies by Dennis Chitty, Charlie Krebs, and their students revealed that the most abundant species was Townsend's vole (Chitty 1967; Krebs 1979; Taitt et al. 1981; Taitt and Krebs 1983; Boonstra 1977; Taitt 1984). Its diet is mostly shoots and roots of grasses and herbs reached via underground burrows and runways through the grass. Populations of voles fluctuate seasonally from fewer than 100 to over 1,000 voles per hectare depending on the intensity of predation in early spring and the abundance of food (Taitt et al. 1981; Taitt and Krebs 1983). These small mammals might also play an important role in grasslands by cycling nutrients and aerating the soil.

Clearly, the heron's food web is a study in complexity. The most likely route of dioxin contamination in the marine ecosystem is a detritus-invertebrate-fish-heron pathway. As more studies were made in the waters near the Crofton mill, it was soon discovered that clams and crabs were also contaminated with dioxins, and this contamination subsequently led to a closure of the fishery. New regulations were established, and the pulp mill industry reduced the levels of contamination. Lower levels of contamination have been detected in heron eggs in recent years, and the fishery has opened once more.

Study Sites

From this basic understanding of how food webs worked, I embarked on a detailed study of the ecology of the great blue heron. The first step was to choose a study site. There were several colonies in the Fraser River delta, but these herons were known to carry some contaminants. I wanted a site that was far from contamination. About 100 pairs of herons nested on Sidney Island in the southern Strait of Georgia in the 1980s, and many of them foraged in an eelgrass meadow in a nearby lagoon. A few eggs were collected and found to have low levels of contamination. In autumn, the herons left Sidney and dispersed among the Gulf Islands, where most were difficult to find. So I spent most of the autumn, winter, and spring studying herons on the Fraser

Abandoned grasslands that harbour meadow voles, such as this one on Westham Island, are important to adults and especially juvenile herons as foraging habitat in autumn and winter. *(Photograph by Rob Butler)*

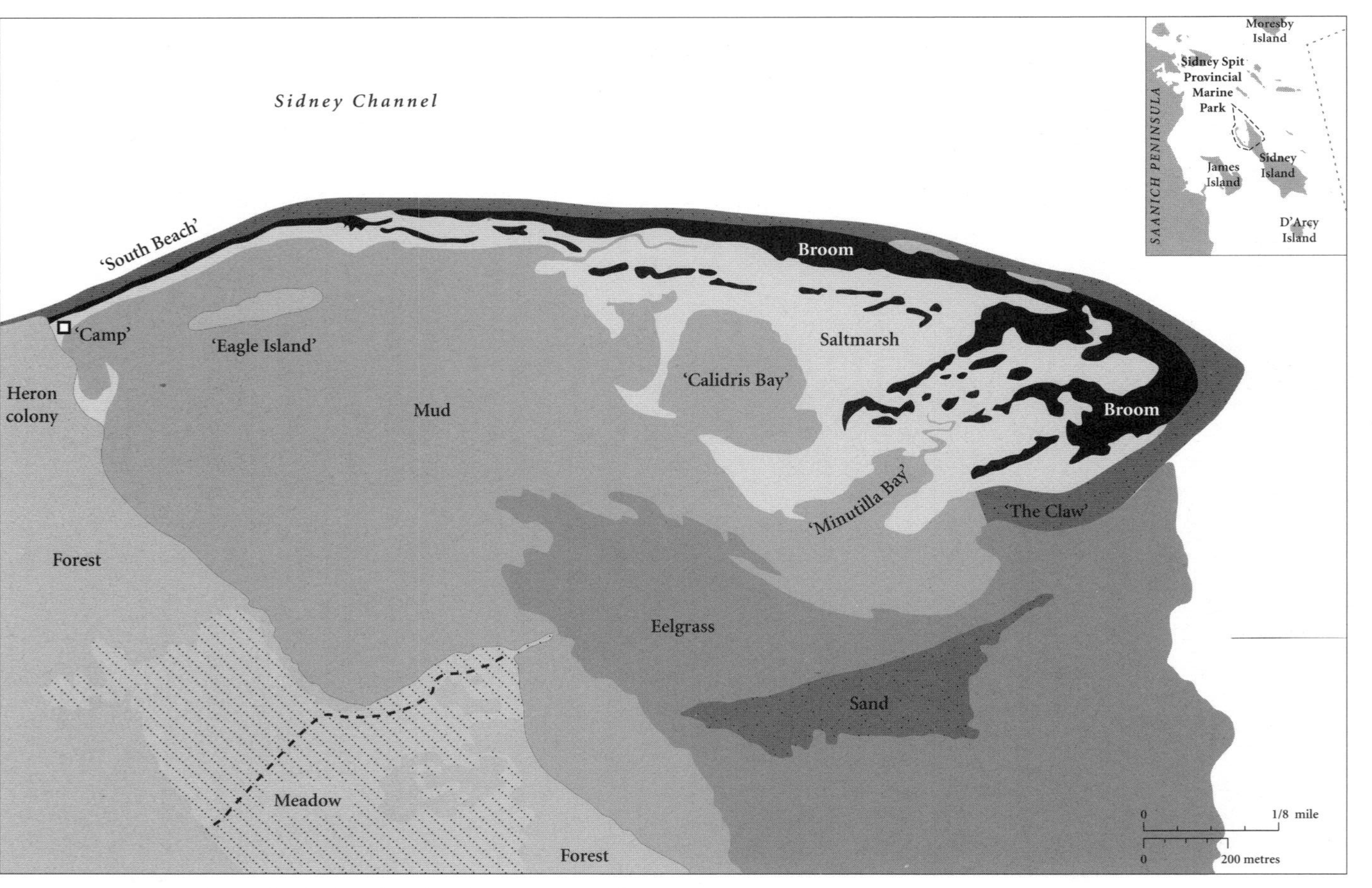

Figure 5 **Lagoon on Sidney Island where breeding herons were studied**

River delta, where herons are a common feature of the landscape. I also made periodic visits to study herons at colonies around the Strait of Georgia.

Sidney Island

Sidney Island is located at the southern entrance to the Strait of Georgia, twenty kilometres northeast of Victoria and four kilometres east of the town of Sidney (see Figure 5). It is just over nine kilometres long on an axis that runs east-west and less than a kilometre wide at its widest point. Sidney Island stands apart from most islands in British Columbia in having wide sand beaches and a large lagoon instead of narrow, steep, rocky beaches. From the air, the island vaguely resembles a swimming otter. The southeast end is rounded like an otter's snout, and much of the island's nine-kilometre length resembles a sinewy body. A 1.5-kilometre spit tapering tail-like into Sidney Channel completes the picture.

The bedrock at Sidney is mostly sandstone protected by an igneous intrusion along the east-end shore. Wind and ocean waves have gnawed near-vertical cliffs rising about thirty metres above the beach along the northern and southern shores. These cliffs are the source of sand and gravel transported by ocean currents to form a ninety-one-hectare lagoon near the northwest end of the island. It was in this lagoon that most herons foraged and that I spent most of my time studying them (see Figure 5).

Reaching to the west, the spit is a narrow strip of sand that ends in a sandy islet that can be reached on foot during low tides. Pilings driven into the sand provide roost perches for cormorants, and the beach is a roost for hundreds of glaucous-winged, California, Heermans, and mew gulls. Shorebirds such as western and least sandpipers, sanderling, dunlin, black oystercatcher, and occasionally ruddy turnstone roost and feed along the spit. Shelving to the north is a wide sand beach where brant, surf scoters, white-winged scoters, and herons feed.

The lagoon is a particularly rich source of food and attracts much of the birdlife on Sidney Island. At its entrance, a sandy hook we called the claw is used as a roost by gulls digesting meals of mudshrimp, sea worms, and fish. Behind the claw and within the lagoon is Minutilla Bay, after the scientific name for the least sandpiper, which finds it especially attractive during migration. Landward of the claw and forming the ridge of Minutilla Bay is a sand dune on which the yellow sand verbena blooms each summer. Along the southern shore of the lagoon, wide swards of glasswort blanket the mud, and

Most herons nesting on Sidney Island foraged in a lagoon formed by wind- and wave-eroded sand from the sandstone cliffs transported to the west by ocean currents. *(Photograph by Rob Butler)*

Unlike most locations on the south coast of British Columbia, the saltmarsh in Sidney Lagoon has not been greatly altered by humans. Herons gather there to await the ebbing tide to expose the eelgrass meadows.
(Photograph by Rob Butler)

on drier land are wide meadows of spike grass and large-headed spike sedge. These dry grasslands, which bake to an ochre brown in the summer heat, are the nesting grounds for the savannah sparrow, which serenades the thickets during the day, and the common nighthawk, which calls from the sky at twilight. Along high dune ridges, scotch broom and wild rye cut green slices through the otherwise drab meadows. Calidris Bay on the southern shore of the lagoon is a favourite haunt for summer migrant sandpipers.

The birds that share the lagoon mudflat with the great blue heron during low tide include the western sandpiper, least sandpiper, semipalmated plover, black-bellied plover, glaucous-winged gull, California gull, mew gull, greater yellowlegs, short-billed dowitcher, and long-billed dowitcher. The western sandpiper is by far the most numerous shorebird. Several hundred birds were present each day in late April and early May during the northward migration; in July and August, when they migrated south, their numbers swelled on some days to over 1,000 individuals. Our banding program showed that individuals stay for only a few days before departing to distant lands (Butler, Kaiser, and Smith 1987). The international nature of the migration of the western sandpiper is evident from sightings of our marked birds in Kansas and Panama (Butler et al. 1996). Mew gulls, California gulls, and glaucous-winged gulls patrol the beds of sea lettuce and eelgrass in search of stranded fish, crabs, and unsuspecting sea worms. Mew gulls are most abundant in March and April, when 100 to 150 were counted in the lagoon. Their numbers dwindle in summer when most are on their breeding grounds in Alaska. California gulls breed in the Canadian prairies and return to the coast in July and August, when numbers in the lagoon peaked at about 100 birds. Between 150 and 1,000 glaucous-winged gulls are present in the lagoon on most low tides throughout the spring and summer.

Deep within the lagoon where the mud is fine grained stands Eagle Island, an igneous rock about one hectare in area. The birth of the lagoon can be traced to the effects of Eagle Island thousands of years ago. Sand eroded from the cliffs of Sidney Island and, pushed westward by the ocean currents, ran up against Eagle Island, first creating a land bridge at low tide and eventually surrounding the small island entirely. Behind the land bridge, the calmed waters began to settle out fine sediments that eventually became Sidney Lagoon.

The history of humans on Sidney Island has been one of destruction and exploitation. Before European settlers interfered with the Tsongees and Saanich peoples, Sidney Island was a coveted source of fish, clams, crabs, and camas bulbs. Native people left behind garbage heaps known as middens in the Gulf Islands that date to 3500 BC, and the size of middens on Sidney Island indicates long use. Shells of the butter clam, little neck clam, cockle, and blue mussel are abundant in the middens at Sidney. In addition to harvesting intertidal animals, Native people burned tracts of forest to maintain flower meadows where the beautiful blue-flowered camas lily would grow. The small bulb provided a form of starch that could be stored throughout the winter. Sidney Island was one of the better sites on the coast because of its dry forests and sunny climate. But burning the forest meadows would pale in comparison to what lay ahead.

In the late eighteenth century, Spanish and English explorers arrived in search of the fabled Northwest Passage between the Atlantic and Pacific Oceans. They charted the Gulf Islands and left their names to many of them. A century of exploration and limited settlement by Europeans followed until gold was discovered in the interior of British Columbia in the 1850s. Hudson's Bay Company records of the time show that Sidney Island retained its Native name of Sallas Island, but by 1860 it was renamed Sidney Island by Captain Richards, who plied the nearby waters aboard the HMS *Plumper.* European settlement spread along the south coast, and by the turn of the century, the Native people had moved to reserves on Vancouver Island, leaving Sidney Island and the surrounding lands for European settlement. Sidney Island soon faced a wave of forest destruction that spread along the south coast of British Columbia. About a century ago, the majestic forests of Sidney Island were levelled and dumped into the lagoon in huge booms destined for mills near Victoria and the Fraser River. Not much remained as a reminder of the original forest during my stay. However, hidden in the shadows of second-growth fir were one-metre-diameter stumps of Douglas-fir spaced tens of metres apart. Hand-axed holes chopped by loggers for their 'springboards' to raise them above the flaring butt of the trees have outlived their makers. A few large monarchs of the past that had been spared the axe towered above the nesting colony of herons. Soon after the razing of the forest, Sidney Island Brickworks mined the clay soil to make bricks for many local landmarks, such as the Empress Hotel in Victoria and Centennial Pier in Vancouver. Shards of red brick littered the entrance to the lagoon, where they were discarded decades ago.

The land title changed hands several times, which brought more changes to Sidney Island. The Todd family in Victoria once owned much of the island

For centuries, Native people harvested flower bulbs, fished, and dug clams on Sidney Island (formerly Sallas Island). Soon after Europeans arrived on the coast of British Columbia, the forests were felled and homesteads created meadows and introduced many exotic plants and animals. The largest meadow on the island lies adjacent to Sidney Lagoon. *(Photograph by Rob Butler)*

and introduced exotic birds, mammals, and amphibians, most of which were gone during my stay. Fallow deer introduced to the island about 1930 denuded much of its herbaceous growth and browsed the lower limbs of trees. Gone were the fields of camas – only a few holdouts were found in rock crevices out of reach of deer. Little understorey vegetation survived the 1,500 or so deer that roamed the island during my studies. In summer, small herds crashed their way through the forest by day and ventured into the saltmarsh at dusk. A few decades ago, the eastern end of the island was purchased by a consortium known as the Sallas Forest Company, which cut and replanted the forests. The value and beauty of Sidney Island was eventually recognized with a formal designation by the British Columbia Ministry of Environment, Lands and Parks of the sand spit, lagoon, and a portion of the forested uplands as Sidney Spit Provincial Marine Park.

Sidney Island lies at the gateway between the estuarine conditions of the Strait of Georgia and the maritime conditions of the open Pacific. The differences in salinity are reflected in the plants and animals that live in the surrounding waters. One of the most conspicuous groups of animals during my stay was the birds. The dabbling ducks found in great numbers in the Fraser River estuary are replaced by seabirds such as auklets, gulls, murrelets, guillemots, and sea ducks near Sidney. The waters that surround Sidney Island run through a network of channels between hundreds of islands. Many regions of the British Columbia coast attract large numbers of fish-eating birds, mostly in winter, but Sidney Channel is remarkable in the number of birds it attracts throughout the year. Red-necked grebes, double-crested cormorants, long-tailed ducks, and surf scoters are abundant in autumn, winter, and spring. In summer, large flocks of gulls, alcids, and cormorants assemble in the channel to feed on a small smelt known as the sandlance. Glaucous-winged gull, California gull, Heerman's gull, rhinoceros auklet, Brandt's cormorant, pelagic cormorant, common murre, and Bonaparte's gull gather in flocks sometimes totalling over 400 birds.

Over the past few decades, the number of nesting bald eagles in the Gulf Islands has greatly increased, from about sixty-five pairs in the mid-1960s and mid-1970s to over 100 pairs in the mid-1980s (Vermeer et al. 1989). On 6 July 1986, a pair began to build a nest on Eagle Island in Sidney Lagoon, and by 26 March 1988, the first clutch was laid. In that year, the pair succeeded in raising one eaglet, and two eaglets were raised in every year except 1995, when no chicks were raised. Eagles are both scavengers and predators. They chase

The waters surrounding Sidney Island support large numbers of fish-eating birds throughout the year. About 1,320 pairs of double-crested cormorants nested on nearby Mandarte Island in 1989 during my study, making it the largest colony in British Columbia. *(Photograph by Rob Butler)*

cormorants, herons, gulls, and ducks in the lagoon and carry dead fish and birds, including heron chicks, to their nests.

The Fraser River Delta

Forty kilometres to the northeast of Sidney Island, the Fraser River slides into the Pacific across a wide delta in southwest British Columbia. The burgeoning human population of Vancouver spilling into the delta has erased much of the delta's former wildness, but a glimpse of what the delta was like a century ago is contained in sketches by the Royal Engineers between 1859 and 1890 (see Figure 6). Margaret North and Jan Teversham (1984) plotted these surveys onto a map with a reconstruction of the shoreline as it was prior to diking. The resulting map reveals a mosaic of habitats shaped by periodic inundation of flood waters. When I read North and Teversham's paper, I imagined the northwesterly winds blowing off the sandheads across the grasslands and whistling in the boughs of giant sitka spruce. White blossoms of crabapples outlined beach ridges stretching toward the North Shore mountains. The sun warmed the sloughs and a flock of herons gathered there out of the wind's reach. Farther out on the wide delta and beyond the great marshes lay mudflats where huge flocks of ducks and geese fighting the northwesterly sweeping down the Strait of Georgia took refuge in the river and marsh channels.

Although the land has seen many changes, the assembly of birds on the Fraser River delta is still a remarkable spectacle. Until a few years ago, there was no comprehensive census of all waterbirds on the delta, so Dick Cannings and I enlisted the enthusiastic birders of the Vancouver Natural History Society to help out. For one year beginning in March 1988, small teams of naturalists censused all birds seen on the ocean side of the dikes around the delta (Butler and Cannings 1989). The dikes were divided into seventeen segments, each segment requiring no more than about two to three hours to census. One day each month closest to the highest tide of that month was chosen for a simultaneous, deltawide census. For twelve months, naturalists counted all birds in their census areas, and by year-end over 1.5 million birds had been tallied! No other estuary in British Columbia came close to the Fraser River delta in terms of number of birds. Shorebirds were the most numerous of all species, accounting for over 34 per cent of the total bird population (see Table 5). About 30,000 dunlin spent the winter on the delta (Butler 1994b). Dabbling ducks (28 per cent) and diving ducks (27 per cent) were nearly as numerous as shorebirds. All other species were far less numerous than

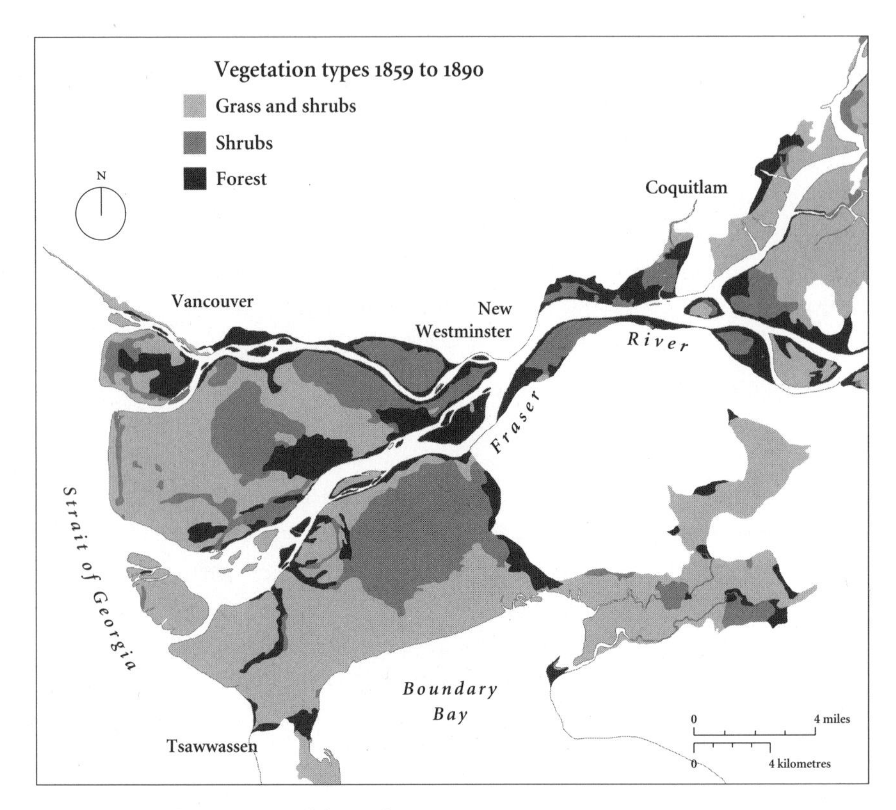

Figure 6 **Map of Fraser River delta in the 1850s**

shorebirds or ducks, but loons, grebes, cormorants, swans, geese, birds of prey, rails, gulls, seabirds, passerines, and herons occurred in substantial numbers (see Table 5). Throughout the year and across census segments, the great blue heron was the most consistently seen of all species.

The remarkable abundance of birds that share the Fraser River delta with the heron is a result of farmlands, extensive intertidal beaches, mild temperatures, and a biologically productive estuary (Lovvorn and Baldwin 1996). Censuses indicated that the greatest densities of most birds occurred in Boundary Bay along the southern shore of the Fraser River delta (Vermeer, Butler, and Morgan 1994). Exceptions were snow geese, trumpeter swans, and herons, which were more numerous at the river mouth. On the landward side of the dikes lies the rich farmland of the Fraser River delta. In autumn and winter, the farmland provides food for tremendous flocks of dabbling ducks, birds of prey, shorebirds, and herons (Butler 1992b; Lovvorn and Baldwin 1996). The important role of farming in the conservation of waterfowl is

The Fraser River delta has internationally important numbers of waterfowl, shorebirds, and herons. Over 1.5 million birds were tallied on its beaches during monthly censuses in 1988 and 1989. *(Photograph by Rob Butler)*

Table 5

Number of birds counted by naturalists once each month in the Fraser River delta in 1988-9*

	Loons	Grebes	Cormor-ants	Herons	Swans	Geese	Dabbling ducks	Diving ducks	Raptors	Rails	Shore-birds	Gulls	Alcids
January	80	362	185	58	270	3,744	66,490	10,059	107	18	93,542	2,942	0
February	108	378	165	60	142	50	13,531	13,484	231	18	14,447	15,302	22
March	147	339	196	170	23	13,537	31,018	54,172	173	31	53,709	12,408	2
April	235	510	168	158	0	2,691	21,449	20,863	41	117	126,687	5,723	5
May	845	1,081	308	202	0	241	1,054	3,805	90	5	3,584	3,027	6
June	2	1	121	182	0	372	654	608	38	0	3,088	3,798	0
July	6	5	179	166	0	428	449	588	45	3	60,456	2,817	0
August	105	1,122	255	293	0	1,300	1,310	10,313	25	7	5,817	15,824	0
September	329	860	330	296	0	396	26,987	13,394	64	13	7,748	11,742	4
October	143	2,346	444	135	19	12,778	100,087	17,496	82	77	101,834	5,257	0
November	200	796	317	186	101	16,071	76,134	97,754	72	124	81,645	4,761	7
December	113	203	245	107	60	11,735	84,021	79,296	184	153	54,575	5,202	2

* Adapted from Butler and Cannings (1989).

clearly evident from the distribution of winter flocks. Jim Lovvorn and John Baldwin (1996) showed that between 75 per cent and 94 per cent of the wigeon, pintail, mallard, and teal in northern Puget Sound and southern Strait of Georgia occur on intertidal habitats adjacent to extensive farmland. Shorebirds roost in fields in autumn. With the return of spring, the ducks that fed on farmland mostly at night in winter switch to foraging there during the day.

Following the arrival of European farmers, the upper marsh was lost to cultivation, so that now only the middle- and lower-elevation marshes persist. The great sedge and bulrush marshes typical of lower and middle elevations in estuarine marshes extend over a kilometre seaward of the dikes in many places. These marsh plants store food energy throughout the winter in underground fleshy rhizomes, from which sprout green shoots in the spring. By summer, the mouth of the Fraser River is fringed by a green swath of marsh about twenty kilometres long from Brunswick Point in the south to Iona Island in the north. The farmers on the delta grow vegetable and berry crops and keep pasture lands for dairy cattle, and they thus provide many of the requirements of large, diverse bird populations. Highways and roads intersect the farmlands, and municipalities grapple with the pressures of urban and industrial sprawl. Areas of conservation have been established or proposed in much of the region outside the dikes, but only remnants of the former habitats on higher ground remain. One of the largest tracts of near-original habitat lies within Burns Bog, which has resisted most attempts to develop its soggy ground. Crabapple trees still grow along dikes on Westham Island, but the vast tracts of grassland are long gone, and only fragments remain of the saltmarshes that once stretched from Boundary Bay to Burns Bog.

The settlement of the Fraser River delta came in a series of waves, beginning with Aboriginal people who, over several centuries, began to exploit the fish, clams, birds, and berries on the delta. Their impact was most noticeable in Burns Bog, where fires were set to prevent the encroachment of pines into areas where blueberries grew. Beginning about 1894, European settlers permanently diked the delta and cleared and cultivated the land. Housing expanded into the delta when bridges were built in the 1950s. In 1985, one year before my study of herons began, about 41 per cent of the delta was in agricultural use, 19 per cent was undeveloped, 18 per cent was in residential use, and about 7 per cent was in industrial use. The remaining 15 per cent was used for parks, roads, and institutions, and less than 1 per cent was secured primarily for wildlife (Butler and Campbell 1987).

The same habitats that support many herons on the Fraser River delta also attract waterfowl. Over three-quarters of the wigeon (shown), pintail, mallard, and teal in northern Puget Sound and the Strait of Georgia are found on beaches adjacent to farmlands during autumn and winter. *(Photograph by Rob Butler)*

In winter, the sedges and rushes in the brackish marshes of the Fraser River delta shed their leaves, which provide food for a web of animals that includes the fish eaten by herons. *(Photograph by Rick McKelvey)*

The Research

The weather on Sidney Island is capricious in spring, making a permanent shelter a necessity. Among a small group of prisoners making amends by working at the CWS office on the Alaksen National Wildlife Area near Vancouver in 1985 was a carpenter intrigued by the challenge I posed to him to construct a cabin that could be transported in pieces by a fourteen-foot Zodiac. Two weeks later, the twenty-five-square-metre cabin was preassembled in a barn at Alaksen. The staff at BC Parks offered to transport the cabin by motorized barge to the lagoon, and a few days later we had a roof over our heads. We were now able to study herons in relative comfort in any month we chose. Over the years, the cabin housed students and visiting researchers from Germany, Mexico, Colombia, New Zealand, England, the United States, and Canada. Their names and quips adorn the inside walls of the cabin, and their records fill the bird record log.

Summer weather on Sidney is more predictable than in spring, so we generally slept in tents or on the beach, ate meals around a picnic table, and used the cabin for cooking and to store our equipment and supplies. The spartan lifestyle kept us outdoors much of the time and in tune with our surroundings. Just watching the goings-on is an important component of fieldwork, and living so much of the time outdoors keeps one in touch with the birds and their surroundings. It is surprising what one discovers in these 'off hours.' While studying crows on Mitlenatch Island, I discovered that oystercatchers do indeed eat oysters, but not very efficiently. John Kirbyson and I documented this fact while eating breakfast on the beach (Butler and Kirbyson 1979).

Over the years, the camp weathered gale-force winds, snowstorms, torrential rains, and an earthquake. Its simple but adequate shelter was welcomed on the many days when winter cyclones slammed the island with high winds and torrential rain. One of those days occurred in February 1988. I had joined Ian Moul and Terry Sullivan to document the behaviour of herons arriving at the Sidney Island colony at the start of the breeding season. The afternoon crossing by inflatable Zodiac soaked us to the skin, and we looked forward to the respite the cabin provided. All evening long, the unrelenting rain slapped against the cabin walls, and high winds howled through the branches of the trees. No herons visited the colony that night. As we cleaned up after a late dinner, the southeast gale momentarily calmed, and the northwesterly gained force a few minutes later. The eye of the storm had passed.

The research on herons was conducted out of the pretentious-sounding 'Sidney Island Research Station,' a small cabin built on the southeast shore of Sidney Lagoon. *(Photograph by Rob Butler)*

The curtain of clouds suddenly lifted, revealing stars of the Milky Way. With the wind at our back and the constellation Orion reflected on the damp beach underfoot, we set out on an evening exploration of the lagoon. The next day, the sun rose in a bright blue sky, and herons waded in the lagoon.

Research on Sidney Island was conducted from 1986 to 1992. From 1986 to 1988, the most effort was spent studying the foraging behaviour of herons in the lagoon. This research included establishing what herons ate and how much food they caught during low tides. In 1988 and 1989, Ian Moul carried out research on the behaviour of nesting herons at Sidney and at the kraft pulp mill near the town of Crofton (Moul 1990). Meanwhile, Darin Bennett began studies of the growth of nestling herons from a small flock of captives raised from eggs hatched at the University of British Columbia (Bennett 1993; Bennett, Whitehead, and Hart 1995). These two studies complemented my field studies to round out the picture of behaviour, development, and ecology.

The sensitivity of herons to disturbance by visitors to some of their colonies was well known, and we wanted to minimize our impact on the Sidney colony. There was little to be gained from studying herons that were nervous about our presence. We learned from other researchers that bird blinds at nest level could result in nest and colony abandonments, especially early in the season. However, very little was documented about the behaviour of herons prior to courtship, so an observation site had to be chosen. Ian's plan was to assemble a blind on the forest floor at a good viewing point on the edge of the colony. Since the alder forest in which the nests were built would not have leaves until April, Ian constructed a covered walkway from black polyethylene sheets suspended by ropes between trees. The walkway was over fifty metres long when finished and allowed him to enter and leave the blind unnoticed by the herons. The blind had a bench that sat three people and a shelf on which to rest elbows while peering through binoculars. The windows were screened with three layers of screen door mesh to block the herons' view of us. I was surprised at how well the herons could hear us moving in the blind or lifting the screen to get a better look. However, they did not abandon nests near the blind, so we were assured that we had taken the proper precautions.

To estimate the reproductive success of herons required repeated visits to nests, visits that neither the herons nor the mature alder trees in which they nested could withstand. In 1986, Ian Prestt paid a visit to the colony on Sidney.

Ian was a pioneer of heron research in England, where he had devised a telescoping pole with a mirror to look into heron nests. After his visit, he sent me a set of poles with a mirror, but I found the device uncontrollable when extended above fifteen metres. Phil Whitehead was also interested in viewing heron nests and wondered if a permanent mirror could be installed above nests. We were worried that the mirror might result in herons abandoning nests. As a solution, Phil fashioned fishing wire, a convex mirror, and an aluminum frying pan into a device that could be attached above heron nests and controlled from the ground. His concept was that the frying pan would serve to cover the mirror when not in use and that it could be removed from the mirror when in use with the pull of wires extending down the nest tree to the ground. We later concluded that the cover was unnecessary. His order of 150 mirrors and frying pans left staff in the government finance department scratching their heads. Not all nests had mirrors installed above them, some pans refused to budge, and other mirrors tilted out of place, but many worked well. At other nests, we resorted to detective work to estimate when eggs were laid and hatched. From nests with mirrors, it was clear that herons threw hatched eggshells from nests, and this 'housecleaning' provided a date on which eggs in each nest began to hatch. Darin Bennett confirmed that heron chicks began to chirp within minutes of hatching, providing us with an audible clue to the stage of hatching in the colonies. Heron chicks stood in their nests when they were about three weeks old, and we were thus able to count them through binoculars and telescope. The success of each nest in raising chicks could be estimated by regularly counting heads.

Someone has said that there is no better way to while away an afternoon than to go fishing, so watching herons fishing must be a close second best. Our routine was to watch a heron through a telescope for ten-minute bouts and record how many fish it caught. Herons spend much of their time stalking small fish. It was relatively easy to record the time of strikes and the type of fish caught by herons. We also estimated the length of the fish by comparing it to the heron's bill. I wanted to know about the fish in the lagoon to compare them with the species of fish eaten by herons. To accomplish this, I needed a beach seine net that could be repeatedly hauled through the eelgrass meadow near where herons foraged. The beach seine net I used was about one metre wide and fifteen metres long. The mesh was six millimetres across. Strung along one long side of the net were floats; along the opposite side, a number of lead weights were sewn into the net so that it formed a vertical mesh fence in the water. The object of beach seining is to catch fish by dragging the net through the water. One foot is placed at the bottom of the net to hold the lead line close to the ocean floor, and with an assistant at the opposite end of the seine net, the haul begins. With great effort, two people can pull the seine ten metres through the eelgrass meadow in about three minutes. Once ashore, we stretched the net as wide as possible and quickly dropped the fish into pails of water, where they could be weighed, measured, and identified later.

During the non-breeding season, herons vacate colonies for neighbouring shorelines, marshes, and farmlands. It was difficult to see herons in the marshes from the ground, so I resorted to using an airplane. Every two weeks, an assistant and I chartered a Cessna 185 to fly over the intertidal beaches and marshlands on the Fraser River delta. Flights coincided with the lowest midwinter daytime tide of about three metres. But neither I nor authorities looked favourably upon low-elevation flying over inhabited areas, so I resorted to driving an eighty-nine-kilometre route through farmland and along beaches once a week. Although tedious, such repeated observations form the database from which seasonal patterns begin to emerge.

1 (previous page) The great blue heron lives year-round along the entire length of British Columbia, where it has adapted to beaches, marshes, and grasslands. *(Photograph by Tim Fitzharris)*

2 (above) The coastal realm of the great blue heron encompasses the calm fjords and shallow-water seashores of British Columbia. High densities of herons are often found on extensive eelgrass meadows, such as this one near Tofino. *(Photograph by Adrian Dorst)*

3 (below) The Fraser River delta is among Canada's hottest real estate markets. Urban sprawl from Vancouver has encroached on farmland where herons and other wildlife lived in the past. *(Photograph by Dave Smith)*

4 (top) Herons build bulky nests of twigs gathered in nearby trees. *(Photograph by Rob Butler)*

5 (bottom) Most herons in British Columbia nest in alder forests, such as the one I studied on Sidney Island. *(Photograph by Rob Butler)*

6 (facing page) Many herons feed their chicks small fish they catch in eelgrass meadows within a few kilometres of the colony. Although flocks in some places number in the hundreds, most herons forage alone in a methodical, slow-stalking manner. This heron has just spotted the movement of a fish in eelgrass. *(Photograph by Tim Fitzharris)*

7 (below) A heron often immerses its head and breast when it strikes at fish. A vigorous shake removes much of the water. Most fish are tossed into the air and swallowed head-first. However, large sculpins can erect sharp spines making swallowing difficult and dangerous for a heron. These fish are carried ashore and speared on the ground with the sharp bill until the spines relax. *(Photograph by Tim Fitzharris)*

8 (above) Herons are more proficient predators when they hunt in shallow water no deeper than their legs. As a result, herons follow the ebb and flood tides across the beach. Sometimes herons become distracted by a fish long enough that the rising tide submerges their bellies. *(Photograph by Tim Fitzharris)*

9 Few animal species better symbolize the need for conservation of the Strait of Georgia ecosystems than the great blue heron. It resides year-round on the shores, wades on beaches, hunts in grasslands, marshes, and streams, and nests in old-growth forests. *(Watercolour painting by Rob Butler)*

CHAPTER 3
FORAGING, FOOD, AND DIET

THE FOOD EATEN BY ANIMALS plays an important role in the habitats they occupy. The great blue heron is catholic in its diet and thus forages in a wide range of habitats, but mostly it wades for food in shallow water. This habit of feeding on a wide range of aquatic animals has allowed the great blue heron to have a wide distribution in North America (see Figure2). How well a heron catches food has consequences for where it resides, its survival, and its reproductive success. And the diverse diet of the heron requires a suite of specially honed skills.

Small fish such as gunnels, sculpins, and shiner perch are the mainstay of breeding herons in British Columbia. Gunnels resemble eels in shape and occur in large numbers in eelgrass meadows. They are eaten throughout the breeding season. Sculpins are consumed mostly when eggs and large chicks are in nests, and shiner perch are eaten mostly when small chicks are in nests (see Table 6). Sculpins occur in large numbers on mudflats and in eelgrass meadows. Shiner perch are schooling fish that move into eelgrass meadows during high tide. The species of fish most frequently caught in beach seines during the herons' breeding season are also eaten in large numbers by herons. The saddleback gunnel, crescent gunnel, three-spined stickleback, staghorn sculpin, shiner perch, and bay pipefish were prominent in beach seine nets and are thus important in the diet (see Table 7). Herons also eat marine invertebrates such as mud shrimp, isopods, and crabs along the British Columbia coast (Verbeek and Butler 1989). In winter, they augment their diet with small mammals caught in grasslands. The main species are Townsend's vole and wandering shrews (*Sorex vagrans*). A few herons beg for handouts from staff at the Vancouver Aquarium, and in spring herons catch frogs in ditches and ponds.

Table 6

Numbers of each of the main prey species eaten by great blue herons during the 1987-8 breeding season on Sidney Island*

	Courtship		Egg		Small chick		Large chick		Total	
Species	*N*	%	*N*	%	*N*	%	*N*	%	*N*	%
Gunnel	20	41.7	188	40.9	56	34.6	175	65.3	439	46.8
Stickleback	0	0	42	9.1	10	6.2	8	3.0	60	6.4
Sculpin	2	4.2	78	17.0	9	5.5	53	19.8	142	15.1
Shiner perch	0	0	20	4.3	68	42.0	11	4.1	99	10.6
Pipefish	0	0	4	0.9	9	5.5	1	0.4	14	1.5
Tube-snout	11	22.9	1	0.2	0	0	1	0.4	13	1.4
Unknown	15	31.2	127	27.6	10	6.2	19	7.0	171	18.2
Total	48		460		162		268		938	

* Reprinted with permission from the American Ornithologists' Union (*Auk* 110:693-701).

Table 7

Number of fish caught in beach seine hauls and those eaten by adult great blue herons in Sidney Lagoon*

Species	Herons	Seines
Gunnels	439	4,627
Sticklebacks	60	882
Sculpins	142	452
Shiner perch	99	265
Pipefish	14	359
Total	754	6,585

* Adapted from Butler (1995).

The winter diet of herons in the marshes on the Fraser River delta consists mostly of fish and shrimp. Anne Harfenist watched herons foraging in marshes and on mudflats on Iona and Westham Islands between November 1990 and February 1992 (Harfenist et al. 1995). At both locations, herons mainly caught sticklebacks, sculpins, and flounders (see Table 8). Also eaten were a small number of other species, including shiner perch, eulachon, redside shiner, peamouth chub, and unidentified species of shrimp and smelt. Most of the flounders eaten by herons were over 105 millimetres in length. Sculpins, flounders, and sticklebacks were eaten nearly every month of the year. Eulachon were eaten only in April, when they entered the Fraser River to spawn. Shiner perch and redside shiner were eaten in spring and summer, chub were eaten in summer, and smelt were consumed in late winter and spring.

Table 8

Number of fish eaten by great blue herons foraging in marshes and on mudflats on Westham Island and Iona Island, Fraser River delta, between November 1990 and February 1992*

Species	Iona	Westham	Total	%
Starry flounder	87	101	188	28.4
Three-spined stickleback	109	70	179	27.0
Staghorn sculpin	97	66	163	24.6
Unidentified shrimp	39	0	39	5.9
Shiner perch	30	8	38	5.7
Redside shiner	0	34	34	5.1
Unidentified smelt	6	10	16	2.4
Eulachon	0	4	4	0.6
Peamouth chub	0	2	2	0.3
Total eaten			663	100.0

* Adapted from Harfenist et al. (1995)

Catching Food

I described the great blue heron as a patient predator (Butler 1995) because of its methodical approach when finding a meal. Herons wait motionless for many minutes for a fish or small mammal to move. They also take advantage of rocky shelves, floating kelp, wharves, and boats to access deep water in search of prey. In Vancouver, herons hunt fish at night from lighted wharves. Where herons wade in shallows in search of fish, they gently place each foot into the water, spreading their long toes for support. As they shift their weight from one leg to the other, the hind leg is slowly raised and slid forward in the water. The neck is outstretched at an angle from the water, and the bill is pointed down and forward. When prey is detected, the heron freezes its motion, lowers its head, and focuses its gaze intently in the direction of its intended prey. It brings its body forward and under its head, bending its neck in preparation to strike. A heron has vertebrae in its neck that allow the neck to bend in a characteristic S shape (see Figure 7). With one quick motion, the near-still heron unleashes its strike. The head shoots forward and snatches the prey from the water. Herons seldom spear their prey – nearly all fish are caught between the mandibles. The neck is retracted to lift the prey from the water.

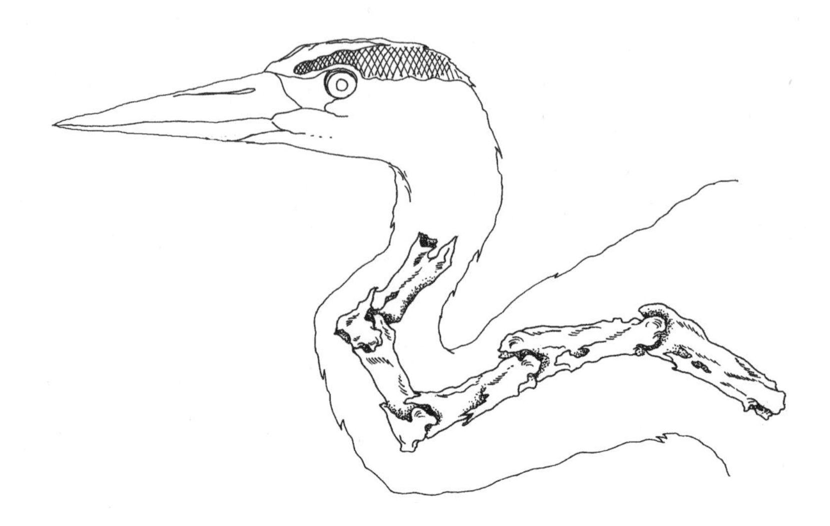

Figure 7 **Neck vertebrae of the great blue heron, shaped to allow a sharp bend creating characteristic S shape**

For much of the time, herons hunt in this slow, methodical manner. Presumably, rapid movements by herons send their prey fleeing out of sight in the depths of the eelgrass. On mudflats, where there is no cover, sculpins and sand dabs scurry away underfoot. Most fish swim five metres or more before settling down. At this distance, refraction of the water and reflection of the sky on the water surface mask their whereabouts. In this situation, herons sometimes run after sculpins and flounders in an unorthodox fashion, with wings partly outstretched.

The immediate response of most fish when caught by a heron is to begin to wriggle. However, the heron's bill has finely serrated edges that provide a good hold. The next step for the heron is to get the fish from the bill into the mouth. Small fish are tossed back into the mouth, but large fish need to be subdued. One begins to appreciate the skills herons possess when they catch large sculpins. Staghorn sculpins lurk in the shallows nipping off clam siphons, eating fish smaller than themselves, and feasting on invertebrates. They are bold fish that possess a nasty set of spines that are erected and locked in place when the fish are captured. And they emit a grunting squeal presumably to startle a captor into releasing them! The spines make swallowing sculpins especially difficult for herons – a large sculpin lodged in the throat could lead to the starvation of a heron. I watched cormorants and mergansers reject sculpins they caught because of the locked spines. Herons deal with particularly tenacious sculpins by spearing them onshore with their sharply pointed bills. This finishing-off takes time away from fishing; sculpins about 130 centimetres in length are swallowed by herons in just over a minute, whereas larger sculpins require over a minute and a half to swallow. Gunnels do not possess the spines of sculpins, but they too can be difficult to subdue when coiled around the bill of a heron. Small gunnels are swallowed in an average of about eighteen seconds, but large gunnels often take several minutes to eat. Adult shiner perch have no spines and cannot coil around the bill, so they are swallowed in less than half a minute.

Contending with wriggling, spiny, or large fish has especially serious consequences for young herons. Robin Gutsell (1995) studied the behaviour of foraging adult and juvenile herons to determine which feeding behaviours are most successful. She found that juvenile herons strike at, miss, and drop prey more often than adult herons. Handling slippery, wriggling fish takes a great deal of practice acquired over many months. Even yearling herons drop a larger proportion of their prey than adults. In addition to a low level of foraging competence, juvenile herons either avoid or simply cannot catch large fish. As a result, they are unable to find as much food as adults. Gutsell also watched herons catch small mammals in grasslands on the Fraser River delta in autumn and winter. Juveniles centred their activity in winter on catching voles. Most old meadows had from one to as many as twenty-five herons present during the day. How the heron locates a vole is not clear, for voles are remarkably fast along their runways. Rough-legged hawks specialize on a diet of voles by plunging from above. Perhaps the underground lives of voles preclude good vision beyond a few metres, but voles can move exceedingly fast.

Herons catch a fish about every two minutes when they have chicks to feed. However, there is much variation in their performance. Some individuals catch several fish in rapid succession and then go many minutes without catching a fish. Herons also catch more prey on ebb tides and in deep water than on flood tides and in shallow water (Forbes and Simpson 1985; Simpson 1984). How many prey are caught is also dictated by the number of fish in the shallows (Butler 1995; Simpson 1984). When fish are plentiful, herons encounter them more often, and their catch rates increase. Some birds increase their foraging rates, but it is unknown if herons do so. Another important factor affecting how much prey a heron catches is the duration of the low tide. In June, the low tide lasts for over 200 minutes on average, compared with about 150 minutes in April (Butler 1993).

Patiently, each heron pursues its prey, with little interaction between neighbours. However, this relative serenity changes when herons pursue schools of shiner perch. As many as 100 herons gathered to feast on such schools in May and June 1987 and 1988 during my studies on Sidney Island, and I have seen flocks exceeding 300 herons catching shiner perch in eelgrass meadows on the Fraser River delta each spring since then. Most of these flocks lasted fewer than twenty minutes. Occasionally, they formed when double-crested cormorants drove schools of shiner perch within reach of herons standing in shallow water. In these instances, flocks of herons followed the cormorants for long periods. But it was the interaction between individuals that was most interesting. Herons, angered by the close proximity of other herons, showed their displeasure by erecting plumes and crown feathers, extending necks and opening wings, and fighting. Their uncharacteristic squawks and displays attracted other herons to the bonanza. Jabbing with the bill and croaking loudly, displaced birds flew a few metres away and began feeding on shiner perch again. These melees were sources of easily gotten food. A deep-bodied shiner perch has nearly ten times as much food energy as a slender gunnel. As a result, a heron can meet its food requirements in slightly more than one hour feeding on shiner perch, whereas it would need the entire low-tide foraging period if it fed solely on gunnels. Herons are presumably willing to tolerate squabbling flock mates because of the higher energy yield from shiner perch.

These energy considerations might be important in explaining the reproductive success of herons. Herons nesting in small colonies are much more variable in the number of chicks they raise than pairs in large colonies (Butler et al. 1996). If nesting success is largely a result of the presence of shiner perch in heron diets, then colonies associated with reliable sources of shiner perch should have the most consistent nesting success. We know little about the natural history of shiner perch. However, the few case studies indicate that shiner perch leave deep water in spring to enter eelgrass meadows, where they mate and give birth (Wiebe 1968). It is during this period of the shiner perch life cycle that they fall prey to herons. Therefore, a shiner perch's choice of which eelgrass meadow to enter depends on the trade-off between reproduction and survival. Pregnant shiner perch have distended abdomens that likely make them less manoeuvrable and therefore more vulnerable to herons than females that are not pregnant and males, while recently born shiner perch are easy prey for herons. The survival of young shiner perch is probably dependent on their rate of growth – small perch are slower swimmers than large perch. Therefore, a pregnant shiner perch should seek eelgrass meadows that provide food and shelter for young shiner perch, while avoiding being caught by a heron. One way to reduce the chances of being caught is to join large schools of other pregnant females. By doing so, an individual reduces the chances of falling victim to a heron. However, the advantage of joining a school diminishes after some point. When schools become very large, the dilution effect becomes trivial. At some point, large schools should break into new schools to pioneer unoccupied eelgrass meadows. Thus, in years when schools of shiner perch are small, females might assemble in a few schools off large beaches; in years when shiner perch are numerous, they should break into many schools off large and small beaches. As a consequence, prey reliability and heron nesting success would be more consistent among colonies associated with large beaches than among colonies near less dependable small beaches.

The Tidal Window

Much of what herons do is centred on catching food. Their daily and seasonal travels, the time of year they breed, and their survival are largely dependent on finding food. Tides set the 'window of opportunity' for herons foraging on fish in British Columbia. The window is open the least in December, when less than one-tenth of the low tide occurs during the day. In contrast, the window is open the widest in May and June, when more than seven-tenths of the low tide coincides with daytime. Thus, the tidal window alone allows herons to catch more fish in summer than in winter (Butler 1993). Prey is also most abundant in summer. Consequently, the food available to herons in spring becomes a signal to them to breed. They move to the beaches in large numbers in March to take advantage of newly arrived fish and lengthening periods of low daytime tides. The number of small fish that enter the intertidal beaches in spring is staggering. I estimated that the number of fish in thirty-one hectares of eelgrass on Sidney Island was over 1.3 million in April (Butler 1995). A few weeks later, in May, over 12 million fish were present, but by month's end the number had fallen to about 1.5 million (Butler 1995). Most of the fish arriving in eelgrass meadows on Sidney Island were gunnels, sculpins, sticklebacks, shiner perch, and pipefish (see Table 9).

Within the seasonal waves of fish species arriving in eelgrass meadows are the comings and goings of different age and sex classes. The first sticklebacks to arrive in April were male. One month later, the number of males had

Table 9

Number of major fish species caught in beach seines at Sidney Island each month, 1987, 1988*

Species	April	May	June	July
Gunnels	18	141	92	73
Sticklebacks	27	98	42	12
Sculpins	21	47	66	58
Shiner perch	8	15	119	50
Pipefish	7	118	31	31
Tube-snouts	5	0	0	0

* Adapted from Butler (1995).

dropped to 43.7 per cent of the 181 sticklebacks, and females predominated in catches in June (52.37 per cent female, $n = 132$). Presumably, this change in sex ratio reflected the arrival of males in April to establish and defend nesting territories in the eelgrass before the fertile females arrived in May. Like the stickleback, male shiner perch preceded females in the lagoon. In April and May, shiner perch were scarce in the lagoon – only ten were caught, and eight of them were male. However, the catches held 57.1 per cent males in June ($n = 182$), when thousands of females arrived to mate and lay eggs. By the end of June, we caught recently born shiner perch, and at the end of July, the six adults we caught were male. The sex ratio of pipefish was about even in May (47.7 per cent male, $n = 88$) and predominantly male in June (69.9 per cent male, $n = 42$). Female pipefish transfer their eggs to a pouch on the abdomen of males. The pouch splits open weeks later, releasing tiny larval pipefish into the sea.

Two fundamental goals of my study were to establish the habitats herons occupy and the role of food in determining when they breed. These questions had implications for the time of year and age of breeding, the number of young raised by heron parents, survival of age classes, sources of contamination, and conservation efforts. It had been theorized that most birds cannot lay their eggs early enough in the season to match the food demands of their chicks to the availability of prey to their parents (Perrins and Birkhead 1983). Prey consumption had been measured in many birds, but measuring prey availability had proven difficult. Prey availability is the number of prey that can be caught where a predator is hunting for food. Fish that are too small or too large to catch, or in deep water out of reach of a heron, are not 'available.' I thought that herons offered an opportunity to estimate both prey consumption and availability with relative ease. I envisioned comparing the

The Canadian Wildlife Service embarked on research of great blue herons in part because of concern for the possible effects of contamination on wildlife in the Strait of Georgia. Deformities, such as this crossed-bill cormorant chick found on Mandarte Island in 1989, are associated with regions of high contamination, although cause and effect are uncertain. *(Photograph by Rob Butler)*

number of fish caught by herons to the number I caught in beach seines. By controlling for daily feeding time using tide heights, I could compare the number consumed to the number present on the beaches through the nesting season. On paper, this seemed a reasonable approach, but in reality it proved a daunting task. The diversity of prey eaten, and their seasonal nature and patchy distribution, required many different techniques, each with its own associated error. And the physical exertion required to beach seine during low tides was enormous.

I began with the easy part: estimating what herons eat. Herons hold a fish in their bills for a few seconds, allowing it to be identified – but estimating how much they eat tests one's patience and fortitude. Some individuals catch fish in rapid succession, and then a lull follows. I look back on the long hours spent watching herons catch fish as an enjoyable period in my life. Perched in the shade on a prominent location overlooking spectacular scenery, the smell of the sea and arbutus leaves carried on a warm breeze, gulls and shorebirds busily feeding along the mudflat – it was hard to beat. Once I had estimated the rate at which the herons ate fish, I needed to estimate the number of minutes the herons foraged each day. Since foraging is dictated by the tides, I reasoned that if I knew the tide when the herons began and ended foraging in the eelgrass meadows, I could then estimate the number of minutes available to them for foraging each day. The Tidal Office at the Institute of Ocean Sciences in Sidney generated these data from their computer model of tidal movements in the Gulf Islands.

The final step turned out to be the most daunting – I needed to estimate the number of fish in the lagoon. My method was simple – I would make repeated seine net hauls in the eelgrass meadow throughout the breeding season, estimate the area I sampled, and extrapolate the density of fish from the sampled area to the entire eelgrass meadow. This method was reasonable if the error in sampling was not too great, but methods of sampling intertidal populations of fish are full of assumptions, including that all fish have an equal chance of being caught and that the sampled area is representative of the larger area. These assumptions made me uncomfortable.

I began by building a fence of galvanized chicken wire sandwiched between two sheets of polyethylene plastic. With assistance from Terry Sullivan, I selected a patch of eelgrass meadow close to sandbars, where the beach seine could be quickly removed from the water. On the ebbing tide in about one metre of water, we pushed four posts into the eelgrass meadow to form a square measuring nine metres on each side. The wire and polyethylene fence was strung along the outside of the poles, leaving the opening closest to shore. Terry and I quickly pulled the seine into position opposite the open side of the fenced eelgrass meadow. We then pulled the seine to the opening and onto the sandbar. Opening a beach seine is like reaching into Pandora's box – you never know what you will catch. There might be an exceptionally large flounder or sculpin. Sometimes we caught flaccid hooded nudibranchs and red rock crabs. Sometimes the net was teeming with small fish wriggling to escape.

The fish were quickly removed from the seine and placed in a bucket of water. Racing against the falling tide, Terry and I positioned the net once more and repeated the procedure. Each time we counted the number of fish we caught, and once the seines were nearly empty, we rested. In late May, when fish are abundant in the lagoon, Terry and I hauled the seine seven times before we were certain that few fish remained in the fenced area. We collapsed on the beach to rest exhausted muscles. Then began the task of sorting, counting, measuring, and weighing our catches.

The time-consuming and demanding work involved in seining clearly showed that the chances of catching different species of fish differed remarkably. Whereas sculpins were mostly swept up by the first pass of the seine, gunnels were still being caught on our seventh haul (Butler 1995). These data indicated that the number of fish in the lagoon at Sidney was in the millions in late May and that herons consumed less than 2 per cent of these fish (Butler 1995).

Herons in British Columbia are unlikely to deplete fish stocks in eelgrass meadows because of species turnover and replenishment. The species, age class, and sex class composition of fish in the lagoon changed with the season. In other parts of the world, herons have been reported to deplete fish stocks. Jim Kushlan (1976) showed that herons removed about three-quarters of the fish from ponds in the Florida Everglades. However, the hydrology of the Everglades is very different from that of the British Columbia coast. In the Everglades, fish are concentrated in shallow evaporating ponds where they can be caught by herons. The ponds become cut off from outside bodies of water, whereas British Columbia eelgrass meadows remain connected to deeper water.

When a female heron catches enough food to make eggs in spring, she is ready to reproduce. But first she must choose a mate, and this is the topic of the next chapter.

CHAPTER 4

SOCIAL AND TERRITORIAL BEHAVIOUR

SELECTING A MATE is one of the most important decisions an animal makes in its lifetime. The mated pair has to be able to work together for over three months to raise offspring. Thus, a good provider will enhance the chances of raising many young and increase the odds that some offspring will survive to breed. For a brief period of a few weeks each year, the generally docile nature of the heron is replaced with intense bursts of courtship activity. The courtship is enhanced by a show of elaborate feather plumes, intense changes in the colour of bills, legs, and bare skin around the eyes, and a rich repertoire of displays. For the rest of the year, the colours fade, some plumes are shed, and displays become uncommon.

Members of the heron family *Ardeidae* grow plumes specifically for courtship display (Hancock and Kushlan 1984). The lacy plumes worn by egrets were prized by the millinery trade as adornments for hats at the turn of the century. Herons grow pointed lanceolate plumes along the neck and chest and across the back, and they grow filamentous plumes on the breast. Lanceolate and filamentous plumes are longer than they are wide. The barbs interlock along the vane of lanceolate plumes, but they are free in filamentous plumes and create a hairy appearance. Herons in British Columbia grow plumes in winter in preparation for the breeding season, and by February they are in prime condition.

Hancock and Kushlan (1984) believed that plumes provide an important function in species recognition, but this function is doubtful. I prefer the view of Charles Darwin (1871) that elaborate ornamentation worn by some male animals has evolved from female mating preferences. Plumes provide a signal of breeding readiness and perhaps the quality of the carrier. In herons and

Members of the heron family grow long plumes on the breast and back for courting mates during the breeding season. The length of the plumes might be a signal to potential mates of an individual's parental abilities. *(Photograph by Wayne Campbell)*

many other monogamous species, both sexes are ornamented. One such group is the alcids: seabirds that include puffins, auklets, murres, and murrelets. These small- to medium-sized diving birds gather on islands each spring to select mates and reproduce. In the north Pacific, plumed auklets reach the greatest level of ornamentation in the crested auklet and whiskered auklet. The crested auklet has a crest that emerges from the forehead and spills forward over the bill. The whiskered auklet also has a plume reaching forward over the bill, two white plumes that rise above the head and resemble horns, and white plumes from behind and below each eye that reach down the neck. Using realistic models of the crested auklet with different experimentally lengthened plumes, Ian Jones and Fiona Hunter (1993) showed that both sexes prefer to court mates with the longer crests. Although Jones and Hunter showed only weak evidence that ornamentation signals the quality of mates, they provided convincing evidence that both sexes have been involved in the evolution of ornaments.

There is some evidence among herons that plumes signal mate quality. The cattle egret grows reddish brown plumes on the chest, back, and head, and the bill, eyes, lores, and legs become reddish during the courtship period. Well-plumed individuals are more aggressive, occupy better feeding sites, and have lower mortality rates than poorly plumed egrets (Woolfenden et al. 1976). In a study of the breeding behaviour of the cattle egret in Barbados, Elsie Krebs (1991) found that individual egrets chose mates with similar plumes, largely because of female preference for well-plumed males. Better-plumed mates fed the chicks more often and had better success at raising chicks when food was scarce than poorly plumed individuals. Well-plumed males were also more likely to attempt to copulate with females other than their mates, and females were more willing to accept males that were better plumed than their mates. Thus, Krebs's study provided convincing evidence that plumes play a significant role in signalling the quality of mates. Krebs did not know whether the plumes indicated different age classes of egrets, a condition suggested by Milstein, Prestt, and Bell (1970) for the grey heron. What prevents a poor-quality mate from cheating by growing longer plumes or participating in more elaborate displays to fool a prospective good provider into mating? Presumably, there is a cost to such trickery. Growing longer plumes or displaying more vigorously takes energy and attracts more agonistic behaviour from other herons. A deficit must be repaid, and the cost might be excessive for the return.

Another feature that changes during the heron's reproductive season is the colour of the skin and bill. Apparently, the intensity varies with age and sex: Meyerriecks (1960) refers to soft parts being brighter in males than in females, but this brightness is apparently variable (Hancock and Kushlan 1984). The intensity of colour reaches its peak during courtship and quickly fades once the eggs have been laid. I used the brightness of the bill as a clue to the breeding stage of herons in nests. Herons in the Sidney Island colony laid their eggs in April, and a few weeks later their orange-yellow bills had noticeably faded in colour. In early May 1989, the herons abandoned their eggs to ravens and crows when an eagle disturbed the colony. When I visited the colony about one week later, the herons had returned, and some individuals were sitting on eggs in the nests. Their bills had turned bright yellow-orange once again. This change suggests that coloration is under hormonal control centred on the fertile period of the female.

Some species of herons and egrets are reported to be able to momentarily flush their bills with colour (Meyerriecks 1960). Cameron Davidson captured this moment in a vivid photograph of herons displaying at their nests on the shores of Chesapeake Bay (Dolesh 1984: 542-3). The photograph shows the bills of the courting herons to be blood red. Other photographs of the herons away from the colony show that they have yellow bills. The photograph at the nesting site also shows that the legs are reddish, whereas British Columbia herons have yellowish green legs. This suggests that there are geographical differences in the coloration of the bare skin and bills of herons. The role of brightly coloured bare skin in herons is unknown, but among magpies Charles Trost suggested that it might serve for individual recognition (Birkhead 1991). Great blue herons have very different facial markings, which I used to identify individuals – and they likely use them in the same way.

Much research has been done on the courtship displays of herons (including the great blue heron) once they have settled in a colony (Cottrille and Cottrille 1958; Meyerriecks 1960; Mock 1976, 1978, 1979), but far less is known about their arrival at colonies in late winter and early spring. Herons gather in large numbers near colony sites around the Strait of Georgia in late February and early March (Butler 1995), but colony sites located adjacent to foraging areas are used as roost sites by some herons throughout the year. They often assemble in open, exposed locations such as on log booms and fields near colonies. On 20 February 1993, twenty-five herons gathered about one and a half kilometres from a colony site at Stanley Park, and on 24

February 1996, single herons occupied sixteen of eighteen nests in the same colony. Gatherings occurred when tides were high, presumably because females foraged when tides were low.

Herons return to colonies in spring both alone and in groups. In 1987, herons visited nest sites at Sidney Island as a group late in the afternoon. After foraging in the lagoon for much of the day, individuals began to assemble in trees on Eagle Island, about 800 metres from the colony. As dusk descended on the lagoon, the herons flew en masse to the colony, where they snapped their bills and called loudly. When darkness began to fall, the colony went silent. From a distance, I could see ten individuals on nests in the fading twilight, and I suspect that these herons were males prospecting for nesting sites. The following morning, the colony was vacated, and herons foraged along nearby beaches.

The next year, I again documented the arrival times of herons at Sidney Island. Rather than assembling in the afternoon as they had done in 1987, herons arrived individually at the colony near dusk in March from distant sites. Each night for the next few weeks, they arrived in the same fashion. Fewer herons arrived on windy and rainy evenings than on calm and dry evenings. On moonlit nights, their silhouetted forms were clearly visible as they circled the treetops or stood on nests. Each evening, the herons lingered for increasing amounts of time until April 9, after which they remained all day (see Table 10). The first week of April coincided with the time when females began to form eggs (Butler 1993).

March and April are the time of year when herons seek mates at the colonies and when they are most noisy and their displays most elaborate. A courting great blue heron is a sight to behold. Newly moulted body plumes splay out from the neck and chest and spill off the back, and the bill turns yellow-orange. In full display, the plumes and body feathers are erect, the wings are partly open, exposing a black patch near the bend of the wing, and the yellow-orange bill is highlighted by the white crown and the black plumes on the head.

Doug Mock's (1976) studies showed that individual herons exhibit a variety of displays that involve one or more of the following sequences. Unpaired males perform a 'stretch display,' which begins with raising the bill to the vertical while fully extending the neck and erecting the feather plumes. At the same time, the heron emits a very uncharacteristic sound resembling 'goo-goo.' This display shows off the brightly coloured soft parts around the

Table 10

Time of first arrival, number of hours spent in the colony, number of territorial interactions, and number of occupied nests at the Sidney Island colony, 1988

Date	First arrival time (PST)	Time spent by herons in colony (hour)	Number of territorial interactions/hour	Number of nests with adults
17 March	1900	0.7	0	29
21 March	1856	0.3	0	14
22 March	0	0	0	0
23 March	1908	0.8	0	10
24 March	1909	1.2	0	10
25 March	1859	1.5	0.7	8
26 March	1912	1.1	1.8	14
30 March	<1900	1.1	0	15
8 April	1929	2.1	14.2	15
9 April	–	24.0	10.5	49
15 April	–	24.0	2.0	60
16 April	–	24.0	no data	61

Source: Butler (1995).

eyes and bill and is meant to draw the attention of females seeking mates. Variations of the stretch display are widespread among other heron species. The 'snap display' begins with the erection of neck and head feathers. The head is moved forward and down, and the display culminates with a snap of the bill when the neck is fully extended. Stretch and snap displays are confined to the nest or the branches of nest trees, but herons also display while in flight. In the 'circle flight display,' a heron laboriously flaps or glides with its neck outstretched. The 'landing display' is used throughout the breeding season rather than being confined to periods of courting. It includes a loud croak upon landing and erection of feathers upon alighting at the nest. This display is probably used as a means of recognition by the other member of the pair. There are many other displays, all of which show off the plumes, bill, and soft parts. Once females are ready to breed, they fly to nearby colonies, where the males are waiting. The first herons to arrive in the colonies in spring are likely males prepared to breed. At this time, individuals settle on nests from previous years or build new nests several days before pairing takes place.

Solitary female herons enter the colonies in search of suitable males, but males are selective too. Mock (1976) found that prospecting female great blue herons are often chased away by the aggressive approach of males. Gradually, females get closer to males by approaching along a branch near the nest. Both sexes clearly size up one another before settling on a mate. One of the most extraordinary examples of mate selection among herons occurs in cattle egrets (Blaker 1969; Lancaster 1970). Female cattle egrets attracted by displaying males alight in nearby branches to watch. As many as ten females might gather in the vicinity of a displaying male. Males are aggressive toward females at this stage; the only successful approach is from behind and onto the male's back! Once there, a female attempts to remain on his back and subdue him. If unsuccessful at dislodging the female, he usually submits, and a few hours later the pair bond. Female great blue herons are not as aggressive as cattle egrets, but both sexes spend time making choices. Once mates are chosen, bill duels occur between mates. With feathers held erect, wings partly opened, and necks outstretched, males thrust their bills at females, who return the sally. Not all displays are as aggressive as bill duelling – herons also rapidly click their bill tips together in the direction of their mates.

The three most frequent displays by herons occur when one of the pair returns to the nest after a long absence, when twigs are brought to the nest, and when mates relieve one another at the nest (Mock 1976). The 'greeting ceremony' occurs when an arriving heron calls and lands at the nest. Its mate usually responds with a stretch display, or it might arch its neck or fluff the feathers on its neck. Wiese (1976) showed that the greeting ceremony assists egrets in locating their mates and nests in the colony. Early in the courtship stage, egrets do not recognize their mates or their calls. Thus, repeated calling helps to establish recognition. Twigs are important to herons because they require effort to find and break free. Males bring most twigs to the nest, and females place them in the nest. There are three stages when twigs are brought to the nest: when the nest is being constructed, when the eggs are laid, and when the eggs hatch. The last two stages presumably reduce the chances of losing eggs and chicks from the nest. The arrival of a male with a twig often prompts a stretch display by his mate. The pair clap their bill tips, and she places the twig in the nest. Nest relief is met with ceremonious stretch displays and bill clapping. Sometimes the pair preen and snooze before one member leaves the nest.

Soon after mates are selected, heron pairs begin to copulate. Copulation occurs on the nest mostly in the morning and evening so that females can take full advantage of foraging at the midday low tides that they need to produce eggs. Thus, at this time of year, males are mostly confined to nests during much of the day, while females are off fishing. The solitude of the colony is occasionally disturbed by squawking and bill snapping. There are few displays leading up to copulation. However, the number of interactions between herons peaks at this time (see Table 10). During copulation, a female leans forward, bends her legs slightly, and partly extends her wings for balance. The male places one foot in the middle of her back and hooks his toes around her partially opened wings. He lowers himself while flapping his wings for balance and grasps her neck in his bill. She moves her tail aside and positions his cloaca against hers. The act lasts for only a few seconds, after which he steps aside. She stands and shakes herself, and both usually begin to preen their feathers.

Herons form monogamous pairs during a breeding season but choose new mates each year. Keith Simpson (1984) followed marked individual herons between breeding seasons at Pender Harbour on the east coast of the Strait of Georgia. Of eighteen banded pairs of herons, none were together the following year. Of twenty-one herons in the colony in 1978, thirteen were on different nests, seven were not seen, and one was on the same nest the following year. All of the thirteen marked individuals that failed to raise any young in 1978 were absent from the Pender Harbour colony in 1979.

Although herons form monogamous pairs, male herons display to neighbouring females when their mates are away. The intention of these displays is for males to copulate with females other than their mates. In a colony in Louisiana, Wiese (1976) noted male great egrets mating with any female that landed on their nests and with females at neighbouring nests. Most of these matings were by newly mated males and neighbouring incubating females. These attempts might be misdirected, or females might store the sperm in case a second nesting attempt is required.

Territory

Herons defend the territory around their nests to reduce the loss of twigs to neighbouring herons and probably to minimize copulation attempts by neighbouring males. However, male herons face a dilemma: they have to feed away from the colony, and doing so allows other males to display to and copulate with unguarded mates in their absence. Copulation attempts from nearby males are uncommon, but they pose a threat to paternity. It is assumed, therefore, that cuckoldry is rare in great blue herons, although the opportunities are available. This is an area where more work is needed. If territorial behaviour around the nest reduces copulation attempts, then defence should be the greatest when females are fertile. Early in the season, herons defend an area of about two metres around their nests. After the eggs are laid, the defended area decreases to about one metre. Consequently, latecomers settle closer to occupied nests as the season progresses. Young, inexperienced breeding herons might nest later to avoid cuckolding attempts by older herons. The hormones needed during reproduction are different from hormones needed for parental care of eggs and young. Territorial behaviour is under control of testosterone, whereas incubation behaviour is controlled by prolactin. If the fertility of male herons declines soon after eggs are laid as a result of increased prolactin levels, then they will pose little paternity threat to recently arrived herons preparing to breed.

Herons also defend good foraging sites by threatening approaching herons with displays, chases, and fights. The delicate choreographed displays of a threatening heron are exquisite. The neck is outstretched, and the bill is held above the horizontal. The wings are partly extended so that they arch out from the body, with the wingtips nearly touching the ground. The tail is angled upward, exposing the white undertail feathers, and the plumage is bristlingly erect. The antagonists approach each other with slow, deliberate steps. Usually, one heron falters and turns away until it is far enough that it can relax its threatening pose. If the threats escalate, a brief skirmish of lunges followed by a chase of flapping wings ensues.

Herons might display ownership of territory by perching on exposed locations. Adult herons often perch on pilings or in trees in their territories. These sites are probably chosen for their safety, but they also might serve as display posts of ownership of the territory. I have seen herons begin threatening displays and launch attacking flights from these perches. Although adult herons shed plumes shortly after the eggs are laid, they maintain many of their contrasting feathers throughout the year. They prominently show black plumes on the head, a white crown, and the black patch at the bend of the folded wing. Some plumes are retained along the neck, breast, and back. These features make it easy to distinguish an adult from the more drab juvenile. An adult heron perched on an exposed river piling is like a beacon to other herons. To my knowledge, no studies of the importance of plumage to territorial herons have been made, but I suspect that herons retain their distinctive plumage to advertise their territorial status.

There is much to learn about territorial behaviour in herons. Most of the observations have been of unmarked birds, so our knowledge of the phenomenon is sketchy. Range Bayer (1978) found herons defending foraging territories along Oregon beaches after the breeding season. Juvenile herons attempted to defend sites for a few weeks until forced by tides to abandon them. Adults defended beach sites for longer periods of time. Adult herons on the Fraser River delta defended riverbank territories, especially in marshes (Butler 1995). Nine territories that I could see both day and night were occupied by solitary herons every month of the year. There was some evidence to suggest that the herons on the Fraser River and those along beaches away from Sidney Island were male. However, I have seen herons with short bills (and therefore female) defending territories along ditches in the Fraser River delta. Thus, it is doubtful that only males defend territories. Not all territories are occupied each year. I know of three territories that have not been used since the abandonment of a colony along the Fraser River in 1992. I have also seen new territories established along the Brunette River and around Burnaby Lake, near Vancouver.

Moult and Feather Care

Feathers provide birds with insulation against the elements, allow them to fly, give them distinctive colours, and advertise their parental abilities. As we have seen, herons have evolved spectacular plumes for advertisement. They appear sometime in autumn and grow through the winter. In British Columbia, herons have feathers missing in the wing from June to November. By replacing feathers in sequence, birds retain their ability to fly while undergoing the moult. The moult sequence of herons has been described as chaotic (Milstein, Prestt, and Bell 1970; Siegfried 1971). But recent examination of black-crowned night-herons indicates that they moult up to eight flight feathers in sequence from the inside out, then skip to the outermost primary (Shugart and Rohwer 1996). Many juvenile night-herons arrest their moult, presumably because they have insufficient time. Shugart and Rohwer (1996) proposed that there is an advantage in being able to arrest the moult when there is insufficient time to replace flight feathers between the breeding season and migration. Much more work on the moult of the great blue heron is needed.

Herons care for their feathers by preening them with the bill. Herons also possess a pectinated toenail for scratching the skin and feathers. *(Photograph by Wayne Campbell)*

CHAPTER 5
BREEDING SEASON

THE OPPORTUNITY TO REPRODUCE is a reward for animals that have survived infancy. Once a mate is chosen, breeding birds have to build nests, lay eggs, and raise young. The choices they make will influence how successful they are as parents. One of the most conspicuous features of breeding herons is their colonies, but how they select sites and why they nest in colonies are not well understood. Although a small number of herons nest as isolated pairs, most nest in colonies ranging from a few pairs to many hundreds of individuals (see Appendix 1). In the period 1989-93 in British Columbia, most great blue herons nested in five large colonies in the southern Strait of Georgia, near the mouth of the Fraser River, and in the lower Fraser River valley (see Figure 8). Not much is known about where herons in the lower Fraser River valley forage, but all colonies along the coast are near shallow beaches with abundant prey during the breeding season.

Herons nest on the ground, on human-made structures, in shrubs, and in cacti elsewhere in North America (reviewed in Butler 1992a), but in British Columbia, herons build nests in trees, including alder, cedar, hemlock, Douglas-fir, spruce, hawthorn, and cottonwood. Alder woodlots are used most often. In this case, nests are built in the canopy, so the colony is mostly on a horizontal plane. In large firs and spruces, the nests are often placed on branches up the tree apartment-style. The choice of tree species is unimportant at first glance, but when colonies move sites, they often settle in the same tree species (Kelsall and Simpson 1979).

As a general rule, herons select sites far from human interference. However, there are several instances in which herons have moved to busy

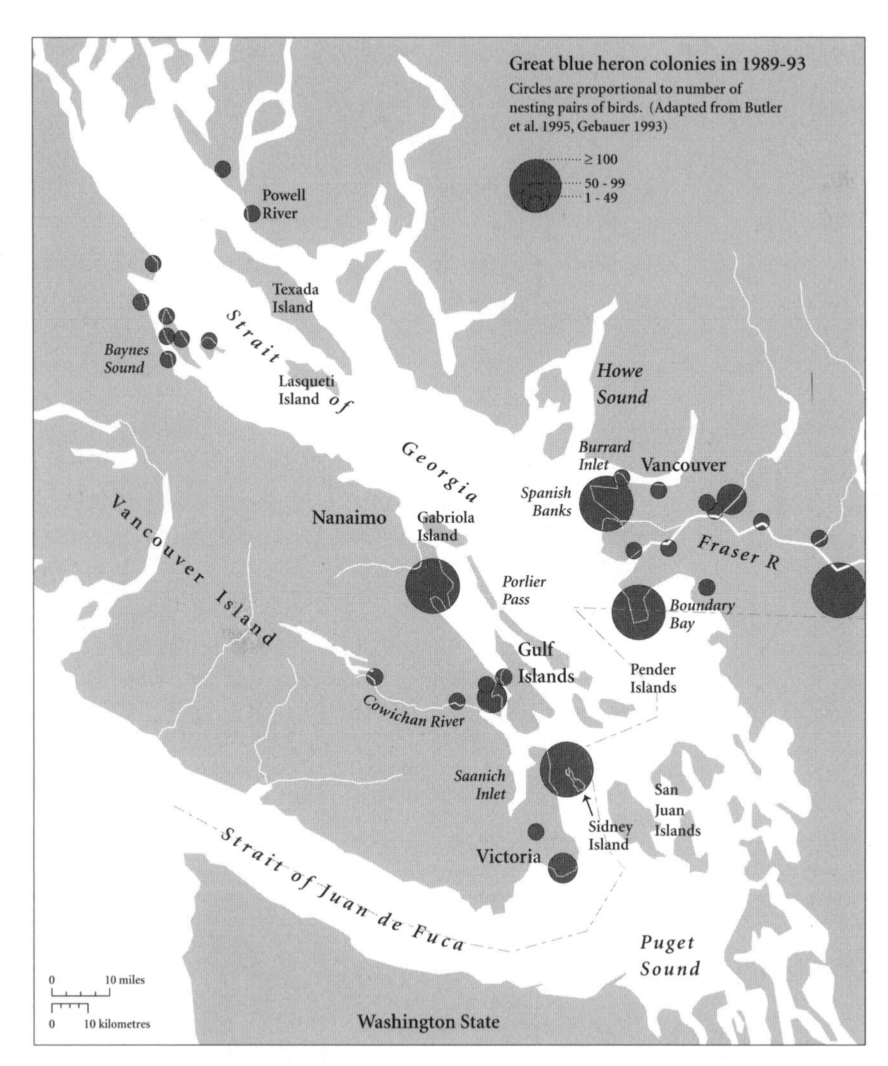

Figure 8 **Distribution of heron colonies in the Strait of Georgia and lower Fraser River valley between 1989 and 1993**

human locations. One of the oldest heronries in British Columbia, in Stanley Park near downtown Vancouver, has been in use since at least 1921. For many years, the herons nested in conifers near Brockton Point, and about 1974 they moved to tall cedars, spruce, and hemlocks within the former Stanley Park Zoo and near the entrance to the Vancouver Aquarium. Amid the hoots and screams of exotic birds and mammals, and the hordes of people below, the herons successfully nested each year. This is the most often viewed heronry in Canada.

The oldest known and most easily observed heron colony in British Columbia is located in Stanley Park. The colony has moved from near Brockton Point, shown here in June 1924, to its present location near the Vancouver Aquarium in the mid-1970s. *(Photograph by D.W. Gillingham by permission of Wayne Campbell)*

The natural history of the heron is explained to hundreds of thousands of visitors each year by aquarium naturalists and on signs belonging to the Stanley Park Zoological Society. Some herons at Stanley Park have learned that they can get handouts of fish from zookeepers tending penguins and from the aquarium marine mammals trainers. An opportunity awaits for a feeding experiment to see if adults receiving handouts begin to nest earlier and raise more offspring than herons that receive no handouts. A second case of herons settling in a busy site involved about twenty pairs that built their nests near the tops of a row of cedars between a hotel parking lot and a busy road to the Vancouver International Airport.

The size of heron colonies has been positively related to the area of the nearby foraging habitat in British Columbia and Maine (Butler 1991; Gibbs 1991; Gibbs et al. 1987). Two jetties about two kilometres long were built on Roberts Bank in the 1960s to provide a terminal for the BC Ferry Corporation and a coal-loading facility. Following construction of these two jetties, an eelgrass meadow began to expand across the sandflat. In the late 1960s, the area of sandflat covered by eelgrass was about 300 hectares, and by 1994 it had expanded to over 400 hectares (see Table 11). Herons began to nest about six kilometres to the southeast on Point Roberts in the early 1970s, and about 200 pairs were nesting there by 1978 (see Table 11). The colony grew over the years, so that over 400 pairs were nesting on Point Roberts by 1992. The eelgrass expanded over the beach at a rate of about 5.4 hectares per year, and the heron colony grew by an average of about ten pairs per year (see Table 11). Such correlations suggest that a relationship exists between the area of foraging habitat and the number of herons nesting nearby, but the causal relationship has not been explained.

Function of Heron Colonies

Why most herons nest in colonies remains a mystery. One view – that colonies provide a central clearing-house for information on where to find good feeding places – has not been well supported (Mock, Lamey, and Thompson 1988). Another view is that colonies of herons might be better able to defend eggs and chicks against predators and thereby raise more young than if they nested alone (Lack 1954). If true, then individual pairs and small colonies should raise fewer young than large colonies, all else being equal. However, isolated pairs and those in small colonies do not raise fewer young than pairs nesting in large colonies (Butler et al. 1995). A third view, and the

Table 11

Area of eelgrass meadow and number of breeding pairs of herons on Point Roberts since the construction of jetties on Roberts Bank in 1960*

Year	Area of eelgrass (ha)	Number of herons
1959	no data	100
1967	348.2	no data
1969	216.0	no data
1975	308.2	no data
1977	no data	183
1978	no data	210
1979	372.2	236
1980	no data	222
1981	432.7	no data
1983	398.3	no data
1987	no data	183
1988	no data	335
1989	no data	256
1990	no data	350
1991	no data	387
1992	420.9	474
1993	no data	400
1994	409.9	414
1995	no data	372

* Data from Appendix 1 and Triton Consultants Ltd. (1995). Regression equations are area of eelgrass = 5.4 year – 10343, $F = 6.8$, $p = 0.04$, $r^2 = 0.53$; number of breeding herons = 7.5 year – 14625, $F = 166$, $p = 0.004$, $r^2 = 0.67$.

one that appeals to me most, is that heron colonies are meeting places for herons at the start of the breeding season (Simpson, Smith, and Kelsall 1987). Herons spend the winter scattered along the coast. As spring approaches, they gather on beaches to feast on the seasonal abundance of small fish that swim into shallow waters to feed and reproduce. These fish provide female herons with the food energy needed to produce eggs. There are many advantages to herons that nest early in the breeding season, but an obvious one is the availability of food. Early nesting pairs get a jump on later nesting pairs by having young in the nest when food is most plentiful on the beach. For this reason, any means of reducing time selecting a mate should be an advantage. Colony nesting might have evolved among herons in the following way. In spring, a female heron must find enough food to make eggs, select a mate, and choose a nest site. She can find large amounts of food by wading in eelgrass meadows, and so these habitats attract large numbers of female herons. Once she has enough food to make eggs, a female heron will seek a nest site and a mate. No time should be wasted since successful nesting declines over time. Thus, she will be attracted to places where many male herons assemble so that she can quickly comparison shop among prospective mates. Comparison shopping is important to females because it saves precious time and energy needed for egg production that would otherwise be expended searching among dispersed males. The comparison-shopping hypothesis might explain why most herons nest in colonies, but it does not explain why some pairs nest individually. To address this question, we need to examine how individuals choose where to nest.

Choosing Where to Nest

While bald eagles hunt herons, the latter do not appear to avoid nesting in the vicinity of eagle nests. Instead, herons seem to prefer to nest near good foraging sites, whether eagles are present or not (Butler 1995). One way a heron might decide in which colony to nest is to copy other herons if nesting success has been consistent between years. In other words, if nesting success depends largely on the availability of food to parents, then future nesting success would be predictable where food supplies are reliable. Thierry Boulinier and I looked at the predictability of nesting success between years among herons in British Columbia (Boulinier and Butler 1997). We discovered that good breeding success in heron colonies in one year does not imply good success in the next year. We interpreted this discrepancy to mean that local feeding areas are unpredictable as sources of food and that it is unlikely herons use breeding success in a colony as a cue to where to breed the following year. However, a good breeding year results in a large increase in the number of breeding herons two years later. This suggests that much of the increase in the number of herons is by recruitment of first-time breeders – herons begin to breed after two years of age. Thus, herons likely choose to nest near rich feeding areas if mates are present.

Nest Size and Construction

Once a heron selects a colony and chooses a mate, it builds a new nest or refurbishes an existing one. Herons seem so out of place perched on tree limbs high above the ground. Even more unusual is their adeptness at finding twigs and building nests in the thin branches of a tree canopy. For all their large size and long legs and necks, herons are surprisingly nimble. Their long toes grip around branches, and their flexible necks counterbalance rocking motions. Herons collect twigs from nearby trees and on the ground near the colony. Most of the twigs are brought to the nest one at a time by males and placed in position by females. Nests from previous years are refurbished with new twigs each year, and nests left unattended are dismantled by neighbouring herons. Nests can be assembled in about three days, but new nests begun early in the season are built over several weeks.

The main structure of the nest is made from twigs the thickness of a pencil and about twenty to thirty centimetres long. Four nests that I disassembled weighed between 2.5 and 4.8 kilograms and represented scores of collecting trips by the herons. However, this range does not encompass the full range of nests used by herons. I could see the eggs through one hastily built nest, while others were over fifteen centimetres deep. Additional twigs are brought to the nest when the eggs are laid and again when they hatch. A nest represents a large investment in time and effort, so nests from previous years are quickly occupied and diligently defended early in the breeding season. However, larger nests do not hold eggs earlier in the season or contain more fledglings at the end of the season than smaller nests (ANOVA, $p > 0.05$).

The distance to the nearest neighbouring nest as measured on the ground directly beneath forty-eight nests in the Sidney Island colony was an average of 3.8 metres ($SD = 1.4$ m). This distance was partly dependent on the spacing of alder trees – most trees held one nest, but a few held as many as four. The diameter at breast height of the alder trees in which the Sidney herons built their nests was an average of 81.5 centimetres ($SD = 13.4$ cm, $n = 48$ nests), and herons in the UBC colony nested in alders with an average diameter of 93.4 centimetres ($SD = 19.0$ cm, $n = 57$ nests). These alders were mature trees over twenty-five metres tall. The position of nests in alders at Sidney and UBC were very similar. Thirty-nine nests at Sidney were in the upper canopy, sixteen were on lower limbs, and four were built against the tree trunk. At UBC, forty-two nests were in the upper canopy, twenty were on lower limbs, and fourteen were positioned against the trunk. Herons that built nests in large conifers positioned the nests on limbs up the tree apartment-style, as at Stanley Park. A single arbutus tree in a colony near the town of Crofton held seventeen nests, a record number in British Columbia colonies. In 1997, Martin Gebauer showed me a small colony in which nests were only about four metres above the ground in hawthorn trees.

The Role of Sea Temperature and Food on Time of Breeding

The production of eggs requires large amounts of energy, and female birds get this energy from stored fat or from food obtained while they are laying.

Scientists refer to species that rely largely on stored food as 'capital' egg layers whereas 'income' egg layers, including herons, rely mostly on energy acquired daily during laying (Drent and Daan 1980). These are relative terms, since most birds use energy from both sources. Many factors can influence when food is abundant and hence when birds begin to breed. Egg laying begins for landbirds soon after air temperatures become warm in spring (Kendeigh 1963; Snow 1955), and some marine birds begin to breed when ocean temperatures rise in spring (Gaston 1992). The mechanism for egg production presumably is closely related to the available food. For herons in British Columbia, the time of breeding is dictated by a chain of factors that includes sea temperature, small fish, and tides.

In 1994, I examined the relationships between sea temperature on the Fraser River delta, the arrival of small fish that herons ate in the intertidal shallows, the number of herons on the mudflats, and the date when eggs were laid in the Point Roberts colony. In a nutshell, I hypothesized that sea temperature in eelgrass meadows near the Point Roberts colony would warm when low tides exposed the meadows during the day beginning in March. Eelgrass would respond to warm water by growing new shoots, and fish would migrate inshore to feed in the meadows. Herons would be attracted to the meadows to catch the fish, which would provide the food energy to produce eggs.

In late winter, sea temperature and fish abundance were documented at the principal foraging habitat for the Point Roberts colony between two-kilometre-long jetties used as a ferry terminal and a ship-loading facility on Roberts Bank, on the Fraser River delta. This intertidal area is about 9.3 square kilometres during the lowest tides, and about 4.4 square kilometres of the beach is covered with eelgrass. Sea temperatures were recorded every thirty minutes between 1 February and 19 May 1994 using a submerged temperature recorder (Ryan Tempmeter) anchored to the ground in the eelgrass meadow where most herons foraged and about 200 metres from where I sampled fish using a beach seine net. A team of keen students from Centennial School in Coquitlam, BC, under the able direction of Rod MacVicar and Ruth Foster provided the muscle power required to do the repeated sampling. The seining was done when tides were about one metre on 3, 18, and 30 March, 13 April, and 13 May 1994 to correspond with the tides when herons could wade in eelgrass meadows. Two students on either end of the seine pulled the net through about 100 square metres of eelgrass. At the end of each haul, the bottom of the net was quickly raised out of the water to prevent fish from escaping. Fish were emptied into a bucket of water, the empty seine net was then pulled about fifteen metres away from the site of the previous haul, and the procedure was repeated until ten seine hauls were completed. A second team of students counted the fish in each bucket, measured the length of each fish, and recorded the mass of samples before the fish were released. The next step was completed using a computer in the office. Robin Gutsell, studying heron behaviour on the Fraser River delta (Gutsell 1995), plotted the length of each fish against its mass and derived mathematical equations for each species of fish. Using her equations, I estimated the mass of the total catch each day (see Appendix 2).

In estimating the amount of food energy in the catch each day, I reasoned that the mass of fish, and hence the amount of food energy, should be related to sea temperature. Herons ingest about 75 to 80 per cent of the energy in their fish diet – the balance is excreted in their droppings. Most readers will be familiar with the conventional calorie as a measure of energy in food products, which is the amount of energy required to raise one gram of water from 14.5° to 15.5°C at sea level. We need about 2,000 calories each day to maintain our weight. Many food products are labelled with their food energy content for the diet-conscious consumer. The metric unit of energy (and work) now widely used by scientists is the joule. One calorie is equivalent to slightly more than four joules. A kilojoule, abbreviated as kJ, equals 1,000 joules. An estimate of the amount of energy consumed by herons (ME) was derived from an equation I developed in another study (Butler 1993). This formula is

$$\begin{aligned} ME &= (1\text{-}0.71 \text{ g dry mass})(0.77 \text{ assimilated})(21.3 \text{ kJ/g dry mass} \\ &= 4.76 \text{ kJ/g dry mass.} \end{aligned}$$

This formula assumes that all fish contain 71 per cent water, that herons ingest 77 per cent of their food energy, and that each gram of dry mass of fish contains 21.3 kJ of energy (Butler 1993).

The final step in the study was to estimate the date of hatching in the Point Roberts colony by searching for hatched shells below a sample of fifty nests every one to two days. The date that females began to form eggs was estimated by backdating thirty-seven days from the date of hatching (ten days for egg formation, and twenty-seven days for incubation).

A switch between night and day low tides occurred in late February and early March. Within this brief period, the beaches began to warm for the first time in the year with each low tide (see Figure 9). Sea temperature in the eelgrass meadows was about 5-7.5°C in February and March because low tides mostly fell at night, but with the increasing number of low daytime tides, the sea temperature rose to about 10°C by month's end (see Figure 9). After late March, sea temperatures increased rapidly when most low tides uncovered the beaches for much of the day, and the eelgrass responded by growing new leaves.

As predicted, the abundance of fish increased in the eelgrass meadows throughout spring. Fourteen species of fish were caught in beach seines on the banks between March and May. The most numerous species were three-spined stickleback, bay pipefish, penpoint gunnel, tube snout, staghorn sculpin, and starry flounder, and all of them are eaten by herons. Five species were present on all days: three-spined stickleback, bay pipefish, starry flounder, staghorn sculpin, and tube-snout. These are widespread species on many beaches in British Columbia.

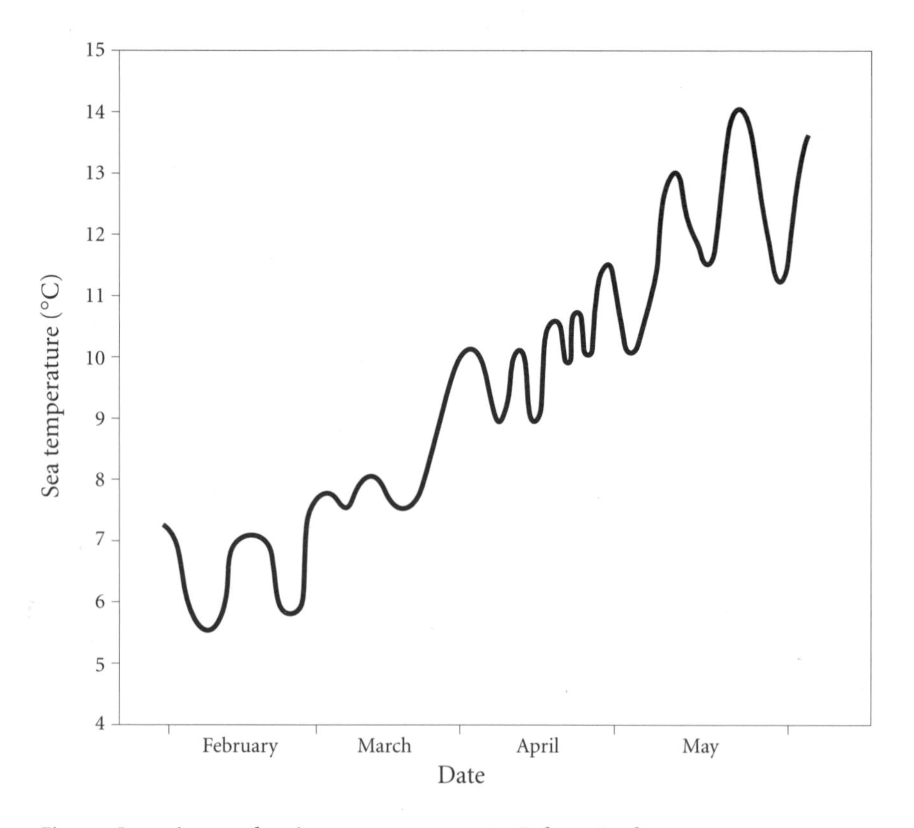

Figure 9 **Late winter and spring sea temperatures on Roberts Bank, 1994**

The daily beach seine catch increased over ninefold from 45 to 419 fish between 3 March and 13 May. The mean (±*SE*) number of fish caught in 10 seine hauls each day was 4.5 (±1.0) on 3 March, 11.3 (±1.8) on 18 March, 15.1 (±2.0) on 30 March, 17.7 (±2.2) on 14 April, and 41.9 (±5.8) on 13 May. When these catches were converted into units of food energy, the estimated amount of metabolized energy (ME) for herons from fish contained in beach seines increased over threefold (1,612.2 to 4,638.6 kJ) between 3 March and 13 May (see Table 12). At the same time, the number of herons foraging on the bank increased during the study from 3 to 196 individuals (see Figure 10). In other words, the number of fish and herons increased as sea temperature on the banks rose (see Figure 11). Heron eggs began to hatch at the Point Roberts colony between 23 and 26 April, reached a peak on 10 May, and declined thereafter (see Table 13). Therefore, by backdating to account for incubation and egg formation, the first female herons began to form eggs on 18 March, and the colony reached a peak of egg production on 3 April. Not surprisingly, sea temperature increased significantly during the study ($p = 0.03$, $r^2 = 0.83$). The respective sea temperature on the banks when female herons were forming eggs between 18 March and 3 April was about 8.6°C and 9.7°C.

Table 12

Number of fish (*n*) and their estimated metabolizable energy (ME, in kilojoules) in ten beach seine nets in eelgrass meadows on Roberts Bank, Fraser River delta, 1994

	3 March		18 March		30 March		14 April		13 May	
Seine	*n*	ME	*n*	ME	*n*	ME	*n*	ME	*n*	ME
1	3	32.6	0	0	9	75.6	24	287.8	41	388.2
2	9	1,169.6	11	71.6	10	143.0	20	161.1	82	795.8
3	8	190.5	13	54.9	11	105.2	13	155.0	56	1,782.7
4	9	70.6	13	90.0	4	74.4	20	162.9	50	301.1
5	3	20.6	23	163.4	14	111.9	12	110.8	31	189.7
6	4	15.1	10	924.0	18	241.4	12	185.3	29	149.0
7	0	0.0	9	63.0	21	188.7	13	163.4	45	270.8
8	1	18.7	14	102.5	19	345.2	16	207.9	14	92.4
9	6	79.4	10	68.0	23	267.8	13	161.9	34	399.9
10	2	15.1	10	68.0	22	608.4	34	368.5	37	269.0
Mean (kJ)	1,612.2		1,605.4		2,161.6		1,964.6		4,638.6	
SE	113.4		85.8		52.0		24.0		158.9	
n	45		113		151		177		419	
Range	0-1,169.6		0-924.0		74.4-608.4		110.8-368.5		92.4-1,782.7	

Table 13

Number of hatched eggs found below fifty nests in the Point Roberts colony, April and May 1994

Date		Number of eggshells	(%)
April	19	0	(0)
	22	0	(0)
	26	3	(2.2)
	29	13	(9.6)
May	3	25	(18.5)
	6	26	(19.3)
	10	29	(21.5)
	13	22	(16.3)
	16	8	(5.9)
	18	9	(6.7)
Total		135	(100.0)

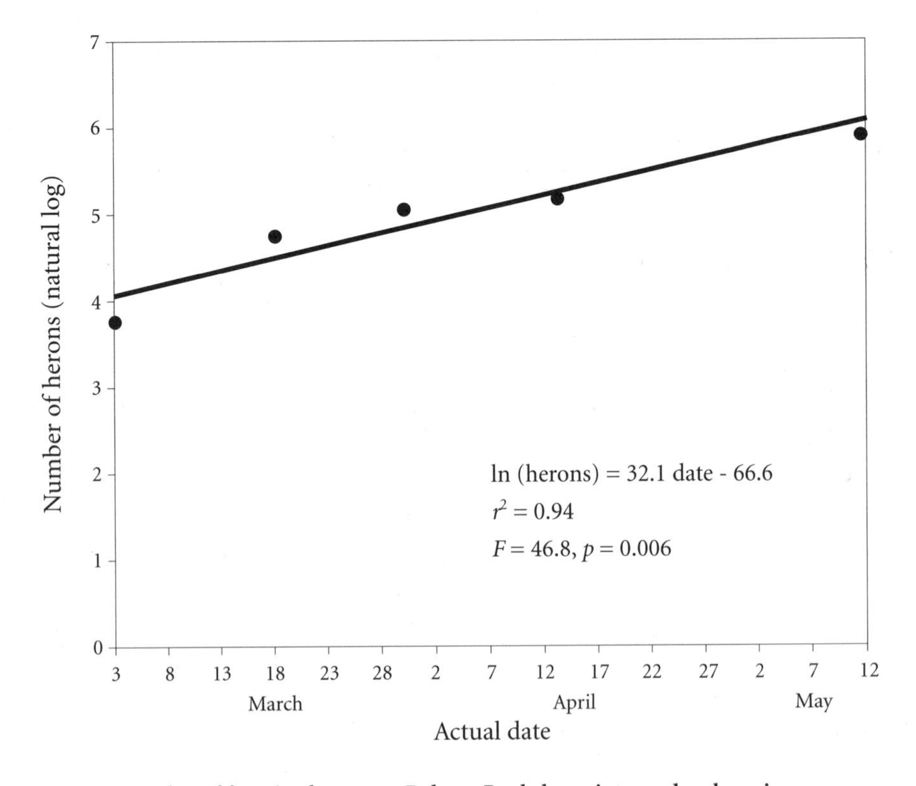

Figure 10 **Number of foraging herons on Roberts Bank, late winter and early spring 1994**

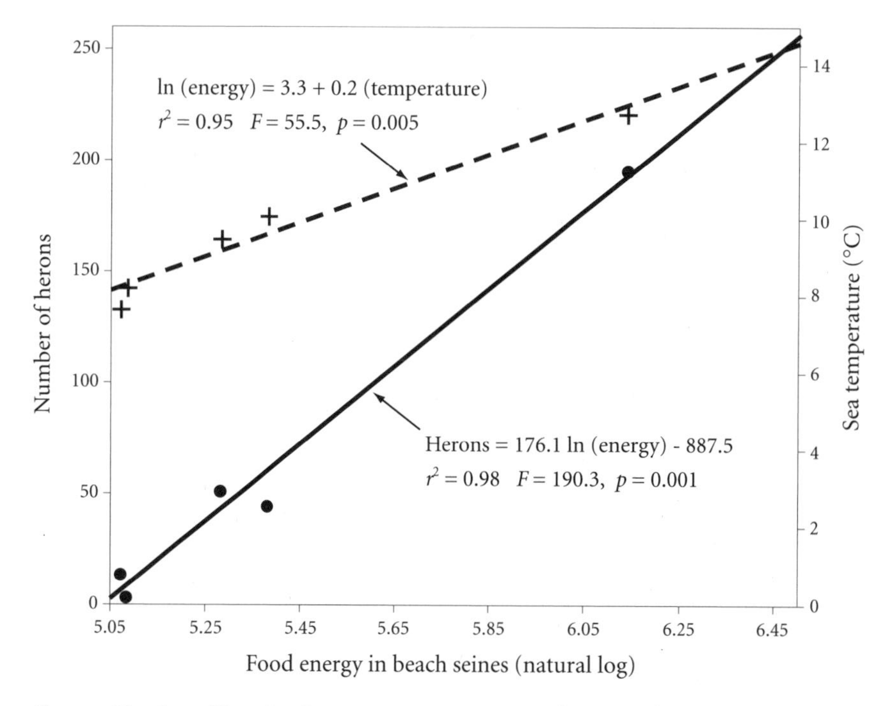

Figure 11 **Number of foraging herons, sea temperature, and estimated energy held by fish on Roberts Bank, March to May 1994**

My findings indicated a relationship between sea temperature and the commencement of egg laying by female herons. The assumption is that herons produce eggs soon after they consume enough fish, and fish migrations are dictated by sea temperature. Scheduling reproduction to the consumption of food energy allows herons to fine-tune their breeding cycle to local food supplies.

These results beg an answer to whether the time of breeding in different colonies is also dictated by the local sea temperature. As any swimmer on the seashore knows, the warmest waters can be found where flood tides cross extensive shallows. During the day, wide sandflats and mudflats lie exposed to the warming rays of the sun, and this warmth is transferred to seawater on the flood tide. On the other hand, steep beaches have less area exposed to the sun, and the flood waters are cooler than in the shallows. As expected, herons in four colonies that foraged in extensive shallows around the Strait of Georgia began to hatch eggs significantly earlier than herons that foraged in deep water (see Table 14). Moreover, herons foraging in shallow habitats were more than twice as successful at raising chicks to fledging age (72.7 per cent versus 31.3 per cent) and thus raised twice as many chicks per nesting attempt (1.6 versus 0.8) as herons foraging along deep-water beaches. This phenomenon provides a simple mechanism to explain how egg laying in colonies around the Strait of Georgia spans several weeks. And it provides a testable hypothesis for future research.

Table 14

Estimated dates that eggs began to hatch in colonies of herons that foraged in shallow- and deep-water habitats

Shallow-water colonies	Estimated first hatch
Tillicum	16 April
Crofton	20 April
Holden Lake	22 April
Little River	11 May
Mean hatch date (*SD*)*	**25 April (11.1 days)**
Deep-water colonies	Estimated first hatch
Hammond Bay	5 May
Denman Island	22 May
Hornby Island	1 June
Powell River	1 June
Mean hatch date (*SD*)*	**22 May (12.7 days)**

* Mean hatch dates are significantly different (one-sided t-test = 5.7, $p < 0.01$, d.f. = 6).

David Lack (1954) thought that birds time egg-laying so that they have chicks in nests when food is most abundant. However, herons seem unable to nest early enough so that the greatest food demands of their growing chicks coincide with the time of year when fish are most abundant and tides are lowest (Butler 1993). Fish are abundant in the lagoon in late May when heron chicks are still very small. More importantly, the food demands of chicks are greatest in June, when they are about three to five weeks of age (Bennett, Whitehead, and Hart 1995) and when food supplies are in decline (Butler 1993, 1995). Clearly, the food demands of heron chicks occur after the peak in food availability to their parents.

As a general rule, birds lay eggs earlier in the year at southern latitudes, presumably because food is available sooner than at northern latitudes. Herons follow this rule (see Table 15). In southern California, herons return to colonies in January and lay eggs in March, whereas herons in Alberta arrive in March and lay eggs in May. Breeding ends for herons in August across most of North America. Therefore, California herons reproduce during about six months of the year, whereas Alberta herons reproduce during about four months. Herons require about three months to complete a breeding season. As a result, herons much farther north than Alberta will barely have enough time to reproduce. In contrast, herons south of California have enough time for two breeding seasons in a year. Studies of breeding herons appear to support this notion. The northern edge of the breeding range of the great blue heron in North America is Prince William Sound, Alaska, at about 60°N (American Ornithologists' Union 1983). In southern Florida (25°N) and on the Galapagos Islands (0°N), herons reproduce throughout much of the year (Powell and Powell 1986; H. Vargas, personal communication).

Table 15

Summary of breeding data for great blue herons from thirteen studies across North America

Location	Latitude (°N)	Number of colonies	Clutch size	Number of fledged chicks: Per active nest	Per successful nest	Nesting success*
Florida	25	26	1.5	1.4	2.0	0.483
California	34	1	2.8	1.5	2.3	0.536
California	38	1	3.2	1.5	2.2	0.469
Nova Scotia	44	4	4.7	2.5	2.9	0.532
Oregon	44	26	4.2	2.0	2.6	0.476
Oregon	45	1	4.5	1.9	2.3	0.466
Quebec	47	22	4.5	1.9	2.3	0.422
South Dakota	44	1	no data	no data	3.0	–
Ontario	48	1	no data	no data	2.1	–
Idaho	48	1	no data	2.0	2.2	–
SE British Columbia	49	1	no data	no data	2.7	–
SW British Columbia	49	31	4.0	1.7	2.5	0.475
Alberta	53	27	5.0	2.3	2.6	0.460

Sources: Blus et al. (1980); Brandman (1976); Butler (1995); Butler et al. (1995); Collazo (1981); DesGranges et al. (1979a, 1979b); Dowd and Flake (1985); Dunn et al. (1985); English (1978); Forbes, unpublished data; Forbes et al. (1985b); Henny and Bethers (1971); McAloney (1973); Powell (1983); Powell and Powell (1986); Pratt and Winkler (1985); Quinney and Smith (1979); Vermeer (1969); Werschkul et al. (1977).

* Nesting success is the mean number of young raised per active nest divided by the mean clutch size.

Eggs and Clutch Size

Most members of the genus *Ardea* lay blue or greenish blue eggs with no markings (Hancock and Kushlan 1984). A sample of great blue heron eggs in North America measured 50.7 to 76.5 millimetres in length and 29.0 to 50.5

millimetres in breadth (Bent 1926), slightly larger than a domestic chicken egg. Twenty-seven freshly laid eggs in a Nova Scotia colony in 1977 had a mass of 70.4 grams, and thirty-four eggs weighed in the same colony in 1978 had a mass of 71.6 grams (Quinney and Smith 1979). Occasionally, very small eggs have been found in nests. One such egg found in a nest on Sidney Island on 11 May 1988 was 44.4 millimetres long and 33.3 millimetres wide. Many species of birds that lay large eggs relative to their mass have correspondingly long incubation periods and briefer nestling periods than birds that lay small eggs relative to their mass (Sealy 1973).

The number of eggs laid by a bird obviously plays an important role in the number of young it will raise. However, the number of eggs that a female lays is dependent on many factors, including when food becomes available to her to produce eggs, her physical condition and genetic makeup, and the time of year. Great blue herons in British Columbia usually lay three to five eggs in a clutch. An eight-egg clutch reported in British Columbia (Campbell et al. 1990) is extraordinary and likely the result of two females laying in the same nest. In forty-seven nests examined at the Crofton colony on 17 April 1986, eleven held three eggs, thirty-one contained four eggs, and five held five eggs, for an average of 3.9 eggs per nest. The relatively small egg (about 3 per cent of female mass) and the two-to-three-day interval between successive eggs suggest that female great blue herons have difficulty finding food and support my earlier contention that herons do not use body reserves to make eggs but rely on the food they consume during egg laying. The great blue heron likely does not lay a replacement egg if one is lost. During egg collections for analysis of toxic chemical contamination, Phil Whitehead found that eggs removed from clutches were not replaced. However, herons will replace entire clutches lost early in the season.

Both members of a mated pair incubate the clutch beginning after the first egg is laid, which results in staggered hatching dates. The incubation period is twenty-six to twenty-seven days in Alberta (Vermeer 1969), twenty-five to twenty-nine days in California (Pratt 1970), and twenty-five to thirty days in Nova Scotia (McAloney 1973). The incubation period is a peaceful time in the colony. While one parent incubates the eggs, its mate is away from the colony or nearby preening, sleeping, or watching. For such a large bird, the heron is able to keep a surprisingly low profile in the nest. The incubation bouts last a long time; males incubate for an average of 10.4 hours and females for 3.5 hours in California (Brandman 1976). About once each hour,

Female great blue herons in British Columbia lay an average of four greenish blue eggs in a clutch. *(Photograph by Rob Butler)*

On rare occasions, herons will lay unusually small eggs. This small egg, shown alongside a normal-sized egg, was found in a nest in the Sidney colony on 11 May 1988. *(Photograph by Rob Butler)*

Male and female herons incubate the eggs in shifts that often last many hours. This pair is tending a nest thirty metres above the ground in a Sitka spruce in Stanley Park. *(Photograph by Rob Butler)*

the incubating parent rises from the nest to rearrange itself, and every two hours it rolls the eggs with its bill. Ian Moul's (1990) observations in the Sidney Island colony showed herons to be very attentive parents. Rarely were nests with eggs left unattended by both parents, and over 90 per cent of the time, one of the parents sat in the nest. In comparison, herons at the Crofton colony incubated eggs for about 85 to 90 per cent of the time, although one adult was present at all times.

Hatching of Eggs and Growth of Chicks

The hatching of eggs brings a renewed level of activity to the otherwise quiet colony. Males begin to collect more twigs to add to the rim of the nest. It takes a chick about two days to become entirely free of the shell and several days before all eggs hatch. In Texas, the first egg hatches an average of 1.6 days before the second, which hatches 1.7 days before the third, which in turn hatches 1.9 days before the fourth (Mock 1978). Hatching is usually closely synchronized between nests in large colonies, and on some days the patter of falling shells sounds like the beginning of a rainstorm in the forest.

Some organochlorine-based contaminants interfere with shell formation in birds, including herons. The most celebrated case has been the thinning of eggshells of birds of prey. Organochlorine contamination of herons in the Strait of Georgia was very low in the 1980s (Whitehead 1989), and climbing to herons' nests to collect an egg for toxic chemical analysis was obtrusive and costly. However, the shell thickness of freshly laid eggs taken from nests and of recently hatched eggshells collected beneath nests at the UBC and Sidney colonies was not significantly different ($p > 0.05$): the average thickness in millimetres of eggshells from nests in 1986 was 0.357 at UBC ($SD = 0.033$, $n = 7$) and 0.384 at Sidney ($SD = 0.013$, $n = 4$), whereas the mean thickness of eggshells collected on the ground was 0.355 at UBC ($SD = 0.019$, $n = 18$) and 0.360 at Sidney ($SD = 0.027$, $n = 9$). These results indicated that recently hatched eggshells collected beneath nests can be used as an unobtrusive technique to measure the thickness of heron eggshells.

A newly hatched heron harks back to its prehistoric ancestry. It weighs about fifty-three grams, wears little down to cover its near-naked pink body, and peers with bluish eyes through small slits. Heron chicks begin to preen their feathers at six days of age, stagger to their feet by two weeks of age, and walk steadily by three weeks of age. By four weeks of age, they begin to flap their wings, and they make short hops to branches near the nest at seven weeks of age. They recognize parents approaching the nest at nine weeks of age, and they begin to leave the nest when they are about sixty days old.

Darin Bennett (1993) raised several herons from eggs at the University of British Columbia, providing a careful study of their behaviour. Within minutes of hatching, the chicks utter a faint 'tik-tik-tik' sound that can be heard over twenty metres away. Chicks sport pale grey down along the back, sides, and head, and especially on the crown. Their bills are stout and pliable. Within two minutes of freeing itself from the egg, one of Darin's chicks had eaten its first meal. The sudden begging and bushy crown probably sent a strong signal to the adults to begin providing the chick with food.

Chick survival is closely associated with the rate of growth relative to siblings. In two months, a fifty-gram helpless heron chick reaches the two-kilogram mass of its parents, develops a two-metre wingspan, and grows a thirteen-centimetre-long bill capable of dispatching fish and small mammals. Bennett showed that fledging male herons are 12 per cent heavier (2,465 grams) than females (2,179 grams) and that both sexes have similar masses as adult herons. The pattern of daily growth of heron chicks is relatively slow during the first ten days and last twenty days in the nest, with most of the rapid growth occurring between about ten and forty days of age. This means that for about four weeks, young herons require a great deal of food to continue their rapid growth. In terms of energy, each chick consumes about 2,000 kilojoules per day between the ages of twenty-six and forty-one days. This is a remarkable demand considering that each adult heron needs about the same amount of energy for its own needs each day and that most parents care for two or three chicks. At the peak period of chick growth, a pair of herons can consume nearly 10,000 kilojoules of energy each day – equivalent to the energy consumption of the average North American! The parents meet the prodigious demands of the chicks by catching large numbers of small fish. Bennett, Whitehead, and Hart (1995) used foraging rates from my studies (Butler 1993) to show that heron parents can supply the needs of two to three chicks, the average brood size in British Columbia.

By far the largest amount of food energy consumed by chicks goes into maintaining body temperature and activity – less than one-third of the energy goes into growth. The timing of when heron chicks can be left on their own at the nest has serious implications for their survival. In the first three weeks of life, heron chicks are incapable of maintaining their body temperature, and their small size makes them vulnerable to predators such as crows

About two to three weeks after hatching, heron chicks no longer need their parents to keep them warm. As a result, both parents leave the nest unattended to find food for their chicks. *(Photograph by Rob Butler)*

and ravens. Consequently, one parent remains at the nest to brood and guard the chicks while its mate goes fishing. On warm and dry days, the parents sometimes stand on the nest, but for much of the time parents cover the chicks to keep them warm. The growing demands for food approach the limit of provision of one parent when the chicks are about three weeks of age, and it is about then that both parents begin to search for food (Bennett, Whitehead, and Hart 1995). This is also the age at which chicks can maintain their body temperature without the warmth provided by the parent; they are also large enough to be safe from most predators. My hunch is that herons that have chicks when food is plentiful in the lagoon can prolong the period of nest guarding and thereby enhance the survival of their chicks over those of later-nesting pairs. Late-nesting pairs might have to prematurely leave their chicks unguarded against predators and the elements because food is not as plentiful in the lagoon as earlier in the nesting season. Thus, parents balance time spent meeting the demands for food against time allocated to brooding the chicks. The maximum food demands are smaller in nests with few chicks than in nests with many chicks. Fewer chicks thus give the parents slightly more time for brooding before both adults have to provide food full time, but the savings are not large. Many parents have more chicks than they can raise in their nests, so the asynchronous hatching of eggs results in the earliest chicks getting a head start on growth over later siblings. Older and consequently larger nestlings dominate their younger siblings by taking the lion's share of the food brought to the nest.

Siblicide

Incubation begins before the clutch is complete among wading birds, birds of prey, many songbirds, and some other bird species. As a result, hatching dates are staggered, and a size hierarchy occurs among the nestlings. Consequently, the last chicks to hatch often die before fledging, grow more slowly, or stay in the nest longer than the first chicks to hatch. The greatest toll falls upon the last-hatched chicks, which often die. Staggered hatching seems unusual, especially when many birds are capable of hatching their eggs in a few hours (Nillsson 1995). The debate among scientists continues about why so many birds hatch their eggs asynchronously, with three general explanations taking centre stage: hatching asynchrony is (1) an adaptation to unpredictable food supplies, (2) a means of saving time and thereby increasing the number of offspring that will eventually reproduce, and (3) simply unusual behaviour (Stenning 1996). The key ideas are that brood reduction (loss of the smallest and weakest nestlings) might occur when food supply and parental effort are insufficient to raise all the chicks, and that parents cannot predict these factors well enough at the time of laying. In such situations, it is less costly to lose chicks earlier and increase the chances that the survivors will be well fed than to postpone the losses and risk losing all the chicks to starvation.

Mock (1985) showed that two-thirds of all heron broods with three and four chicks suffer losses in which the youngest chicks stop growing, presumably because they are receiving insufficient food. Early in life, chicks eat regurgitated food dropped into the nest by parents, but as they grow the young herons scramble to swallow as much as they can right from the mouth of the parent. The oldest chick, being the largest, often gets the greatest share. Mock showed that chicks continue to beg after nearly 90 per cent of all feedings, and he interpreted this behaviour to mean they are not satiated. By eighteen days of age, heron chicks can catch the food as it is dropped by the parents, and by four weeks of age, they can lift and swallow whole fish from the nest. Large fish delivered to the nest result in tug-of-wars between siblings. Mock believed that the number of chicks that survive depends on the amount of food adults can deliver to a nest, but he did not think that mortality is simply mediated by hunger. His argument was based on the fact that fighting in egret nests begins soon after hatching and, even when food is plentiful, does not cease when chicks are satiated. He thought that aggression increases in situations in which heron chicks are fed small prey, because it can be monopolized by the largest siblings.

A similar situation probably applies to herons in British Columbia, where prey sizes are also small and squabbling between siblings is common. The latest research suggests that the degree of aggressive behaviour in chicks is determined by the amount of testosterone deposited in eggs, which declines with laying order (Mock 1996). If food supplies differ between years and broods are adjusted to food supplies, then we should see years when large broods are common and years when they are scarce. Average brood size has fluctuated by as much as one chick at the UBC and Nicomekl colonies and by more than one chick at some small colonies (Butler et al. 1995), but I do not have corresponding data on food supplies.

Entering a heron colony when there are large chicks in nests is an experience that can repulse the uninitiated. The 'guk-guk-guk-guk' banter of the young rings throughout the colony. It is interrupted by the hoarse, loud 'gronks' of arriving adults, and the heightened expectations of a feeding set the young into a frenzy of calling. Nest sanitation is restricted to directing the faeces out of the nest and down the tree to splatter on leaves, branches, and unsuspecting ornithologists below! A hat is a good investment for this job. If the chicks become upset, a regurgitated fish might join the faeces raining down from a nest. Great blue herons return to nests at all hours of night and day. However, the greatest activity occurs during low-tide periods, and the least activity occurs after dark. On calm nights, I often slept on the beach beneath the stars. I welcomed the silence of the night after a day in the colony, though the quiet was periodically disturbed by landing calls ringing from the forest.

Nesting Success and Post-Fledging Care

Clutch size nearly doubles across the latitudinal range of the great blue heron, and about half of all eggs laid become fledged chicks (see Table 15). This analysis indicates that parents are equally efficient across the range of the species. Within a season in British Columbia, the number of fledglings raised per breeding pair declined because of an increased number of total nest failures (Butler 1995). Nesting pairs that laid eggs in the first week of the breeding season on Sidney Island and failed to raise a brood comprised 16 per cent. The failure rate for pairs that laid eggs in the third and fourth weeks was 33 per cent, and for pairs laying eggs in the fifth to eleventh weeks, 60 per cent failed to raise any young (Butler 1995). Grey herons in Belgium also had reduced success as the season progressed (van Vessem and Draulans 1986). I do not know why late-nesting pairs are less successful than early-nesting pairs. Food supplies begin to wane after late May, so perhaps late-nesting pairs are unable to find enough food to continue caring for their young. Late-nesting pairs might also be younger and less experienced parents than early-nesting pairs.

By two months of age, young herons are as large as their parents, which seem reluctant to feed the hungry brood. Adults often alight in branches away from the nest and pause for a few minutes before approaching the young. Once an adult arrives at the nest, the grown chicks lunge at its bill and pull its head into the nest. The adult usually drops the food and quickly departs. Heron chicks become independent of their parents' care soon after leaving the nest. At first, the young herons make short flights to tree limbs near the nest, and gradually they take longer flights within the colony, but they quickly assemble back at the nest when a parent returns with food.

By June, heron chicks in British Columbia are about two months old and as large as their parents. *(Photograph by Wayne Campbell)*

The post-fledging stage of the heron's life is not well understood. Some fledged herons accompany their parents to nests for food for about ten days after leaving the nest. I expected young herons to make inaugural flights to foraging grounds with their parents but was surprised at how few showed up in the lagoon at Sidney Island. In 1988, over 150 young were raised at Sidney, but only about twenty were seen outside the colony at any one time. Puzzled by this low number, I took note of how herons dispersed from the colony. My notebook records describe an incident on 20 July 1988:

> Juvenile settled in tree at 0704h. As an adult departed toward Victoria, the juvenile followed about 15 metres behind calling. After about 0.5 km, the juvenile returned to tree near colony. At 0710h a second adult left in direction of Sidney and the juvenile called while following adult from colony. Another juvenile followed 2 adults toward Cowichan but fell behind about 1 kilometre from colony.

I saw two fledglings flying awkwardly with an adult over my home en route to the Fraser River. These incidents suggest to me that juveniles follow departing adults to foraging sites, which probably explains why few juveniles were seen near the Sidney colony.

Other species of herons disperse rapidly from nesting colonies (Erwin et al. 1996), so it seems reasonable to expect that great blue herons behave similarly. A small number of heron chicks marked at the Nicomekl colony in the Fraser River delta disappeared quickly. One was sighted near the colony, one was found near the town of Gibsons, about sixty kilometres to the northwest, and three were seen in Pitt Meadows, about twenty kilometres to the northeast. None was seen at nearby Boundary Bay. Moreover, recoveries of herons banded at the UBC colony in Vancouver in the 1970s were widely scattered: nine were in the vicinity of the Fraser River delta, three were found in Washington State, three crossed the Coast Range and were found in the interior of British Columbia, and one was located in Oregon (Campbell et al. 1990). Although some great blue herons disperse widely after breeding, not all do so.

Why herons disperse quickly is unknown and raises many questions about which birds disperse and about the advantage of dispersal. The first fledged, the eldest siblings, or both might exclude latecomers from foraging areas near colonies. Bayer (1978) found that herons establish territories around foraging areas in autumn, and I have seen herons at Sidney Island and on the Fraser River delta defending stretches of beach. However, interference between foraging herons is very uncommon (Butler 1995; Gutsell 1995). Perhaps juveniles use the summer to explore other potential foraging areas and future breeding locations. If fledglings mostly follow adults to foraging grounds, which seems likely given the few observed cases, then their dispersal location will depend largely on the movements of adults. The observation of fledglings following parents back to nests to be fed seems to indicate that some stay close to their parents at least for a brief spell after leaving the nest. However, what subsequently happens to the fledglings requires more investigation.

CHAPTER 6

CHOOSING WHERE TO LIVE

One of the most important decisions an animal makes toward its survival is choosing where to live. Herons do not actually ponder the alternatives before acting, but natural selection should favour individuals that exhibit the behaviour that results in the most offspring surviving to become breeding adults. Studies of the behaviour of animals have clearly shown that not all individuals behave the same way. There are many choices to be made by herons, and those choices are influenced by the season, their age, future opportunities, and so on. There are other possible reasons that explain how herons choose where to settle. Dominant individuals might exclude weaker herons from good foraging sites, so that the poorer and usually younger age classes have no choice but to seek marginal areas (Bayer 1978; Richner 1986). Another possibility is that the most proficient individuals can find sufficient food in habitats where other herons would starve (Butler 1994a). However, no matter what the situation, I assume that individual herons choose to live where it is most beneficial to them.

The choices herons make have serious consequences for their survival and ability to reproduce. Herons have to find enough food to survive each day, and if they are to breed, they need to find extra food for egg formation and eventually for their nestlings. Thus, it comes as no surprise that herons nest near rich foraging grounds (Brandman 1976; Dowd and Flake 1985; Pratt 1970; Simpson, Smith, and Kelsall 1987). For young herons, finding enough food for themselves in winter is very time consuming, and many die as a result. Thus, the consequences of poor decisions might be reduced or failed nesting attempts or increased risk of death.

Before I could understand how herons choose sites, I needed a description of the year-round use of habitats by the age and sex classes. This entailed long hours of observation of adults at breeding colonies and of all age classes throughout the winter.

Breeding Herons

Each day as the ebbing tide drained Sidney Lagoon, adult herons left the colony to gather in the saltmarsh and on driftwood logs. On brisk days, they found shelter behind thickets, where they preened feathers, slept, or stood in the warmth of the sun. As the tide ebbed farther, a few individuals ventured into the water to hunt sculpins and flounders on the mudflat, but most waited until the eelgrass meadows could be reached by wading. They would fish for hours, returning to the colony when the tide pushed them from the beach or when they had enough fish to feed their young. The boisterous calling, jostling, and squabbling by chicks being fed by a steady flow of arriving adults when the tide was low fell noticeably quiet when high tides covered the feeding grounds later in the day.

This scenario was played out each day throughout the breeding season with minor differences as the season progressed. However, it was clear that some herons departed the colony during high tide and left the island for sites many kilometres distant. I thus needed to watch herons throughout the day to make any sense of their movements. From a vantage point, I recorded the comings and goings of all herons and their destinations. By alternating sessions of watching with sessions of rest from dawn to dusk over a two-day period, I managed to assemble a complete set of data for a typical day. I repeated these sessions during the breeding season to get a picture of how herons divided their time at the colony and away feeding. These sessions gave me ample time to contemplate rival hypotheses and even ponder the world's problems.

I have chosen two sample periods to represent the daytime pattern of arrivals and departures of adult herons at the Sidney Island colony. The first sample period was 4-5 May 1987, when most herons were incubating full clutches of eggs in their nests, and the second period was 19-21 May 1987, when most of the parents were busiest providing their growing chicks with food (see Figure 12). Herons arrived at and departed from the colony at all hours of the day. However, the pattern of arrivals and departures differed throughout a day and between incubation and chick-rearing periods.

Arrivals were greatest in the morning and evening during both periods (see Figure 12). Departures, on the other hand, were greatest around midday during the incubation period and in the early morning and late afternoon during the chick-rearing period.

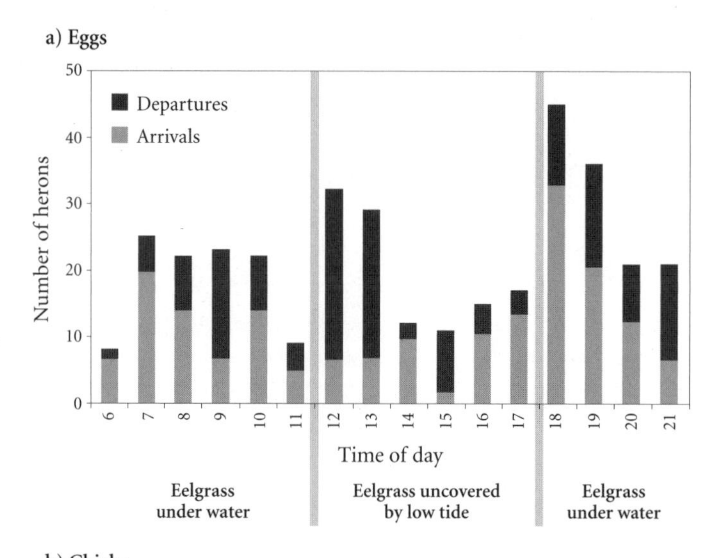

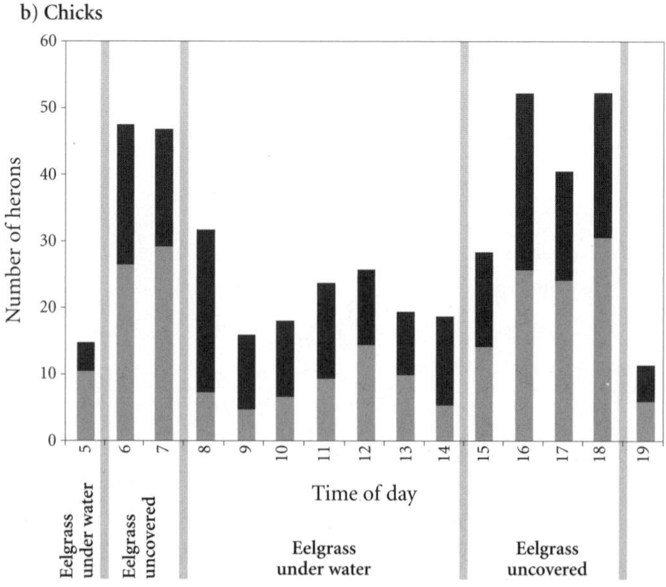

Figure 12 **Arrivals and departures of herons at the Sidney Island colony when eggs and chicks were in nests**

Some of this behaviour could be explained by the position of the low tide in Sidney Lagoon, where most of the herons fed during the day. A large exodus of herons occurred between 1200 and 1300 hours on 4-5 May, corresponding to the time when the eelgrass meadow was first uncovered by the tides. The large number of arrivals between 1800 and 1900 hours on the same day corresponded to the time when the flood tides first made the eelgrass too deep for herons to reach. Similarly, the morning and afternoon periods of activity during chick rearing corresponded with two low-tide periods during the day. First, there was an average of about twenty-three arrivals and departures per hour during the egg stage, compared to over thirty-two arrivals and departures during the chick stage. Second, the number of arrivals versus departures in each hour was much greater during chick rearing than incubation (see Figure 12). The greater activity during the chick-rearing period reflected the emancipation of both parents from brooding young chicks and the heightened task of additional foraging for the chicks. Food supplies were most abundant in late May, and both parents were fully occupied in feeding their chicks. This fact was also reflected in the number of arrivals versus departures each hour. When eggs were in nests, one parent sat in the nest while the other often perched nearby for long spells. But when chicks were two to three weeks of age, both parents rushed to and from the lagoon to take advantage of the short-lived abundance of fish uncovered by ebbing tides. The importance of the lagoon as a foraging site increased during the breeding season up to late June, after which the number of herons in the lagoon began to decline (see Figure 13).

Before long, I began to see a pattern. Female herons generally used the lagoon, while male herons usually flew off the island. I discovered this pattern by watching which herons remained at the colony during the low tide and by looking at the lengths of the bills of herons in the lagoon and off the island (most females have shorter bills than most males; Butler, Breault, and Sullivan 1990). However, the males that flew off the island departed when tides were mostly high, and this puzzled me. Why did they leave the colony to forage when tides were high and at night? To find answers, I plotted the compass bearing of the departures. To my surprise, many of the herons were leaving in the direction of the Cowichan River estuary nearly thirty kilometres to the west! Equipped with a night-vision telescope, Terry Sullivan and I searched for these herons along the shores of the Saanich Peninsula and Cowichan River. Arriving at the Cowichan River estuary near dusk, we found

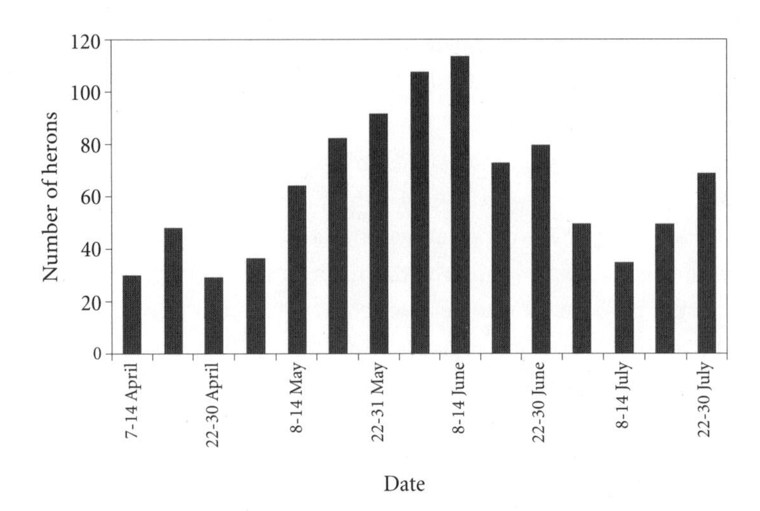

Figure 13 **Maximum number of herons in the Sidney Lagoon during each week of the breeding season**

male herons, but how could we be certain they were from Sidney Island? The only way to find out was to watch the arrivals and departures. Soon it became clear that the herons were indeed arriving from the compass bearing of Sidney. To confirm our suspicions, we searched for herons along the same flight path at intermediate locations and were rewarded. We found many male herons defending territories along the seashore where they foraged in the afternoon and through the night. The following morning, they returned to the colony in time for the females to fly to the lagoon to forage during low tide. In this way, females were away from the colony for most of the day, while males were away for most of the afternoon and night.

Within the colony were some individuals that did not conform to the pattern of behaviour of the masses and instead flew to distant eelgrass meadows as much as ten kilometres from the colony, when other herons foraged nearby. I did not understand why these maverick herons chose to make such long flights when apparently good foraging was close at hand. This behaviour did not seem to be the 'best' choice and in fact might have been a poor choice. None of these herons was marked, but Simpson (1984) showed that herons from a colony at Pender Harbour that flew long distances to forage raised fewer young than herons foraging on the nearby beach. Perhaps these individuals had prior knowledge of good foraging sites off the island, or the distant sites may have been sufficiently important to them at other times of the year that they were compelled to maintain a year-round presence. Or maybe these herons were making poor choices and paying a price.

Very little was known about the whereabouts of yearling herons in summer. Yearlings visited Sidney Island to forage in the lagoon, and one occupied a nest in the colony in June and July 1989, but it did not find a mate. Yearlings were seen with adults in the Cowichan, Goldstream, and Fraser River estuaries and on beaches up to twenty-seven kilometres away from Sidney Island. Therefore, I assume that yearlings reside alongside foraging adults. Similarly, the whereabouts of all the Sidney herons in winter was unclear. Individuals were found scattered in bays and among the islands around Sidney, but I could not account for the 200 breeding adults during searches in winter.

Year-Round Habitat Use

Herons are abundant on the Fraser River delta through the entire year, and – like the herons at Sidney – they can be found near their colonies during the breeding season. In March, herons assemble on beaches during low tides near their summer colonies. The herons from the Nicomekl colony gather in Mud Bay, the Point Roberts colony assembles between the BC Ferry and Westport Terminal jetties, and the UBC colony largely uses the region between the jetties on Iona Island (see Figure 14). By November, great blue herons on the Fraser River delta are in uplands and marshes because of food shortages. Herons spend the halcyon days of summer fishing in eelgrass beds during low tides and dozing or preening feathers in nearby saltmarshes during high tides. However, each day an almost imperceptible change takes place in the eelgrass meadows, and eventually the herons are forced to search elsewhere for food. The tides that fall mostly during the day throughout midsummer occasionally begin to fall near dawn or dusk, and the abundance of fish in the eelgrass meadows in late spring slowly declines. The fish that remain are larger and more difficult for herons to catch. For a heron, the declining foraging opportunities mean that more time is devoted to finding food as autumn approaches. The first herons to feel the pinch are the least proficient foragers. Adult herons are skilled foragers and thus have a greater suite of options on where to forage, a luxury not available to the less efficient juveniles. The only option for juveniles is to seek food in other habitats, and many move to grasslands and ditches on the delta. This shift begins in late August for some juveniles, and by October most of them have left the eelgrass meadows. Many adults continue to forage in eelgrass meadows through November, and a

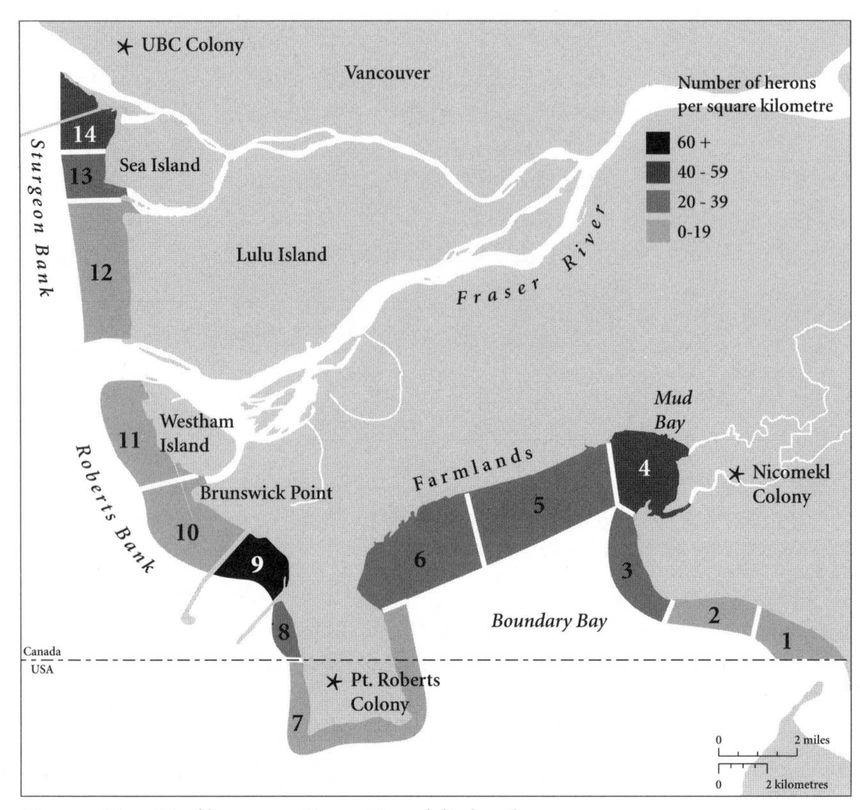

Figure 14 **Density of herons on Fraser River delta beaches**

small number remain through winter, but most move to marshlands at the river mouth where they can wade on most winter days (see Figure 15).

Herons not only choose which beach to forage along – they also choose where on the beach to fish. John Krebs (1974) pondered whether herons share information in some way on the whereabouts of good foraging sites while on the mudflats. Using models of foraging herons placed on the mudflats, he was able to convincingly show that arriving herons were attracted to foraging flocks on the mudflats. I saw similar behaviour when herons changed flight courses and settled with flocks foraging on the mudflats of the Fraser River delta. Herons tended to clump in flocks across the entire delta rather than disperse more evenly across the Fraser River delta (Butler 1995).

Why Herons Move between Habitats

It was clear from my studies that age classes of herons use habitats differently

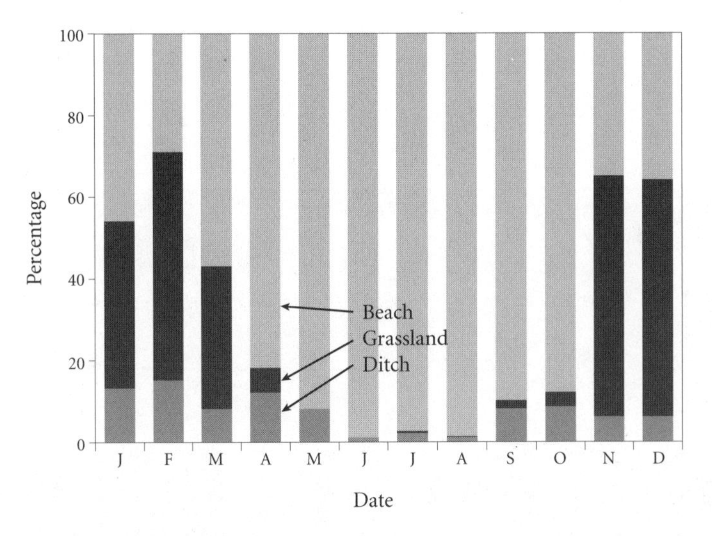

Figure 15 **Percentage of herons on beaches, in grasslands, and in ditches on the Fraser River delta each month, July 1987 to March 1989**

and that juveniles are less efficient fishers than adults (Butler 1995). However, it was a mystery why juveniles that forage alongside adults in eelgrass meadows should avoid marshes where adults forage in winter. My office with the Canadian Wildlife Service in Delta was located in the former estate of George Reifel, and the extensive farmlands, estuarine marshes, and woodlots are a haven for birds and nirvana for an ornithologist. On most days, the staff stroll on the surrounding grounds of the Alaksen National Wildlife Area during their lunch break. During one of these strolls, I discussed the question of how herons selected habitats with Ron Ydenberg from Simon Fraser University. Knowing Ron's interest in foraging theory, I asked him whether he thought that herons might make a good study species to test ideas of foraging risk. Ron was intrigued by the idea because many studies had supported the notion that animals from bats to bumblebees are sensitive to foraging risk (reviewed in Stephens and Krebs 1986). The opportunity to explore this question arose when Robin Gutsell arrived at Simon Fraser University to begin studies with Nico Verbeek. Robin was interested in how animals learn and in the consequences of learning on their survival.

Many animals take risks in their choices of foraging places, and one of the most celebrated examples is that of the yellow-eyed junco (Caraco, Martindale, and Whitham 1980). Juncos are small seed-eating birds that spend much of their time foraging on the ground. Caraco and his coworkers

As autumn approaches, herons begin to leave the eelgrass meadows for the marshes at the mouth of the Fraser River. Many adults will remain in the marshes for the winter, and periodically join juveniles that resort to nearby grasslands. *(Photograph by Wayne Campbell)*

For a few days each year, the British Columbia coast is frozen when arctic air spills over the Coast Mountains. This is a difficult time for herons, which feed on voles in fields. *(Photograph by Rob Butler)*

provided yellow-eyed juncos with two trays of seeds. On both trays were seeds covered by an opaque piece of paper. On one tray, the cover always had the same number of seeds, whereas the other had between zero and ten seeds. Once a seed was taken, the other tray was removed. What Caraco found was that a junco that had eaten enough to meet its energy needs for the day chose the certainty of the tray with the constant number of seeds. However, juncos that were hungry took the risk of finding a bonanza on the tray with varying numbers of seeds. The policy for juncos was clear: be conservative if the daily need had been met, and be risky if not. This policy has been used to explain the habitat choices and foraging behaviours of many animals (Caraco 1981; reviewed in Stephens and Krebs 1986). It was left to Robin to decide if juvenile herons are sensitive to foraging risk.

Terry Quinney's studies in Nova Scotia and my studies in British Columbia had shown that juvenile herons are less successful than adults at catching prey (Butler 1995; Quinney and Smith 1980), but we did not explore how this foraging (in)efficiency might affect a heron's choice of habitat. Thus, Robin Gutsell embarked on a study of herons from summer through winter to coincide with the periods when herons move between habitats. Her findings indicated that in September and October, when the shift from eelgrass meadows to marshes and grasslands begins, adult herons catch larger sculpins than juvenile herons (Gutsell 1995). This difference has a profound effect on the amount of food energy that herons consume. The amount of energy contained in these large sculpins is between about twenty-six and 360 kilojoules depending on the size of the fish. A heron requires about 1,500 kilojoules of energy each day (Butler 1993), and by excluding large sculpins from the diet, juvenile herons put themselves at risk of starving.

Robin then shifted her attention to herons in grasslands. Earlier, I had found that about 95 per cent of juveniles and 82 per cent of adults were found in grasslands. However, fewer than 4 per cent of juveniles and 18 per cent of adults were in marshes in late autumn and winter (Butler 1995). When Robin repeated the survey a few years later, she found nearly identical results. She estimated that juveniles cannot catch enough fish in the marsh to meet their food requirements but that they can meet or exceed their needs by feeding on voles in grasslands on some days. A key finding was that the amount of food consumed by herons is nearly seven times more variable in grasslands than in marshes. Thus, a juvenile heron foraging in grasslands might be sated or go hungry on any winter day, but it would go hungry in marshes on all days. Robin's results were startlingly clear – juvenile herons are at risk of starving and make the best of a bad deal.

Robin also found an unexpected twist in the story. When she began her study, we assumed that juvenile herons are not able to catch enough *small* prey to meet their energy needs, but she found that they cannot catch enough *large* prey. Thus, the most important thing a juvenile heron must learn is how to catch and handle big fish, large sculpins in particular. In fact, their survival depends on it, and that is the subject of the next chapter.

CHAPTER 7

LIFE HISTORY TRAITS AND POPULATION DYNAMICS

In general, animals with long lives breed later in life, have fewer young, and spend more time caring for them than animals with short lives (Goodman 1974; Williams 1966; Zammuto 1986). The great blue heron can live for many years, breeds after two years of age, raises about two young per year, and devotes about three months to the care of eggs and young. It has also adapted to many environments and become the most widespread heron in North America. Equally widespread is the grey heron of Europe, Asia, and north Africa, and it shares many of the same life history traits (see Table 16). One way to examine how the great blue heron has become so successful is to compare it to other species with more restricted ranges. We can also compare British Columbia herons with other populations to understand how they have adapted to life along the British Columbia coast.

No species of heron in North America is seriously threatened with extinction. However, there is some concern about the reddish egret, American bittern, and least bittern. The reddish egret is confined to marine shores of the Gulf of Mexico, whereas the two bittern species breed in freshwater and saltwater marshes. Reddish egrets are vulnerable to human disturbance, while concern for the bitterns focuses on habitat loss resulting from marsh drainage, mostly in cultivated regions of the continent. Thus, the vulnerability of these three species is a result of their relatively restricted habitat requirements compared to the great blue heron. (A similar scenario occurs in Europe, where the purple heron, which nests in reed beds, is more seriously threatened than the widespread grey heron, which occupies many different habitats.)

Table 16

Comparison of biology of the great blue heron and the grey heron

	Great blue heron	Grey heron
Breeding distribution	Southern Canada to Mexico; Galapagos	Europe, central and southern Asia, Africa south of Sahara
Habitat	Lakes, rivers, marshes, swamps, seacoast, grasslands, mudflats	Lakes, rivers, marshes, swamps, seacoast, grasslands, mudflats
Breeding commencement	February (California) to March (Alberta)	February (Great Britain) to May (northern China)
Length of season	110 days	110 days
Organization	Largely colonial, some individual pairs	Largely colonial, some individual pairs
Mixed species	Yes	Yes
Nest	Twig platform	Twig platform
Construction	By female	By female
Nest location	Ground, shrubs, trees, human structures	Ground, shrubs, trees, human structures
Size of eggs	61.3-65.6 mm x 41.9-46.5 mm	61 mm x 43 mm (Europe)
Laying interval	2-3 days	2 or more days
Clutch	1-8; 5 (Alberta), 2.8-3.2 (California), 1.5 (Florida)	1-10; 4-5 (Europe), 3-4 (India), 3 (Madagascar)
Incubation	28 days	25-6 days
Care of young	Both parents	Both parents
Nestling period	65 days	65 days
Brood reduction	Yes	Yes
Age at first breeding	Some at 1 year, most after 2 years	Some at 1 year, most after 2 years

Sources: Hancock and Kushlan (1984); Vermeer (1969); this publication.

The subspecies of great blue heron in British Columbia shares several features common to both southern and northern populations. Like its southern relatives, it disperses after the nesting season within the region where it breeds, but it has a strong seasonal breeding time similar to northern heron populations. It lays a smaller clutch and consequently raises fewer chicks than other herons nesting at this latitude (see Table 15, p. 85). Generally, large-bodied birds devote less of their food energy to making eggs and more time to the care of their young than small-bodied birds. Thus, it is not unexpected that the great blue heron lays the smallest egg for the mass of the female of any North American heron (see Table 17). It also has one of the longest incubation periods – twenty-eight days – among North American herons; only the American bittern exceeds it with an incubation period of twenty-nine days (see Table 17). Many heron species incubate their eggs for only slightly shorter periods of time than the great blue heron, but the largest difference arises in the time spent raising chicks. Great blue heron parents care for their chicks for about sixty days (see Table 17). The American bittern tends its young for about fifty-five days, but they leave the nest at a much earlier age. All other heron species stay in the nest for fewer than about forty-five days (see Table 17). The great blue heron devotes over three months to raising eggs and young, which requires a reliable source of food near the colonies. To raise a large brood requires both parents to provide for the young. This need is met by a monogamous bond for the breeding season and probably explains the elaborate courtship displays, during which prospective mates can size up each other's potential as a parent.

There are few life history data from which to compare the subspecies of great blue herons. Clutch size increases more or less steadily with latitude in North America, from about 2.8 eggs per clutch in southern California to five eggs per clutch in Alberta (see Table 15, p. 85). The relationship can be described mathematically as clutch size = 0.110 x latitude – 0.739 ($r^2 = 0.75$). From this formula, the expected clutch size of herons nesting on the south coast of British Columbia (latitude 49°N) is 4.7 eggs, whereas the actual clutch size is 4.0 eggs. This finding suggests that herons in British Columbia lay smaller clutch sizes than expected of herons nesting at this latitude. Clutches in Nova Scotia heron nests have been more than half an egg larger on average than in British Columbia (McAloney 1973; Quinney and Smith 1979). As a consequence of smaller clutch sizes, the number of chicks raised in British Columbia nests is smaller than that of other herons at similar latitudes (see Table 15, p. 85). These results provide the tantalizing suggestion that British Columbia herons produce smaller clutches than other migratory populations. If true, what feature in the life history of these herons is being traded off, as predicted by life history theory? Do herons that reside year-round in British Columbia live longer as adults than migratory herons nesting elsewhere in Canada?

Table 17

Life history variables of North American herons*

Species	Mean clutch size	Mean clutch mass (g)	Incubation period (days)	Nestling period (days)	Maximum longevity (years)	Female body mass (g)
Great blue heron *Ardea herodias*	2.9	217.5	28	60	23	2,100
Green-backed heron *Butorides virescens*	3.9	72.0	25	35	8	212
Little blue heron *Egretta caerulea*	4.0	108.7	24	40	14	315
Cattle egret *Bubulcus ibis*	3.7	107.2	23	40	17	338
Reddish egret *Egretta rufescens*	3.5	127.1	26	45	12	450
Great egret *Ardea alba*	2.9	145.2	24	42	23	813
Snowy egret *Egretta thula*	4.0	96.4	18	25	11	371
Tricolored heron *Egretta tricolor*	4.5	117.8	21	35	17	334
Black-crowned night-heron *Nycticorax nycticorax*	3.1	126.9	26	42	21	883
American bittern *Botaurus lentiginosus*	4.2	156.5	29	55	8	706

Source: Hancock and Kushlan (1984).

* Scientific names follow American Ornithologists' Union (1983, 1996).

Survival of Eggs, Chicks, Juveniles, and Adults

Estimating the relative survival at different ages of life is a technique widely used by life insurance adjusters. For my purpose, I needed to know when herons suffer the greatest losses and if populations are increasing, decreasing, or stable. To estimate these parameters required estimates of birth and death rates. Birth rates required information from nests, and death rates required following herons throughout the year.

Information on clutch size and hatching success required looking into nests, but climbing to great blue heron nests in treetops in British Columbia was throwing fate to the wind. Even professional climbers shunned away from scaling many trees. Spindly, mature alders seldom grow straight up and are often rotten in places. No one wanted to disturb the herons for long, and the professionals we hired provided the speed we desired. Fortunately, much of the information I needed was gathered without the need to scale trees.

I estimated the number of breeding females at Point Roberts by counting the number of nests used between 1987 and 1991 (see Table 18). I assumed that each female laid an average clutch of four eggs. With the help of professional climbers at the Nicomekl colony, Phil Whitehead estimated that slightly fewer than half of the eggs or chicks were lost during the nesting season. Estimating the number of chicks in nests from the ground was relatively straightforward, with a sore neck being the only occupational hazard. At about three weeks of age, chicks became boisterous and began to clamber about the nest and nearby tree limbs. My method of tallying chicks in nests was to visit the colony after the chicks were well developed but before they began to clamber into tree limbs, about mid-June in British Columbia. Through binoculars, the chicks were counted quite easily. In large colonies, my neck gave out before I could census all nests, so I selected a sample of about fifty nests early in the season to represent the entire colony. About half (47.5 per cent) of the eggs laid by herons became fledglings in the Point Roberts colony (see Table 18). This value was very similar to that for other colonies in North America: the weighted mean nesting success in 139 colonies across Canada and the United States was 0.46 fledglings raised per egg laid (see Table 15, p. 85). Some of the loss was attributable to eggs that failed to hatch – about 15 per cent of ninety-six eggs in twenty-three nests in the Nicomekl colony did not hatch. Corvids took unguarded eggs and left tell-tale puncture marks on the eggs. In 1987, ravens took an estimated 11 per cent of the eggs at Sidney (Butler 1989).

A commonly used method to estimate survival of age classes of birds is to band cohorts, wait for the birds to die, and tally the proportion of individuals that die in each age class. Henny (1972) and Bayer (1981) used this method to estimate the survival of great blue herons in North America. However, this method has many biases that can lead to imprecise estimates of real survival (Lakahani and Newton 1983). The recovery rate is dependent in part on the search rate, which is not even across all heron habitats, and the date that a band is recovered from a dead heron does not always correspond with the date that it died. There are other problems too. Once a cohort has been banded, many years must lapse to allow all birds to die. Newer techniques are available to analyze recapture data from which estimates of survival

Table 18

Estimated egg, chick, and juvenile losses from the Point Roberts colony between 1987 and 1991

	Year					Total
	1987	1988	1989	1990	1991	1987-91
Minimum number of breeding females	183	335	256	350	387	1,511
Number of eggs laid[a]	732	1,340	1,024	1,400	1,548	6,044
Number of eggs and chicks lost[b]	311	670	538	665	697	2,881
% lost/eggs laid	42.5	50.0	52.5	47.5	45.0	47.5
Number of fledged chicks	421	670	486	735	851	3,163
% fledged/eggs laid	57.5	50.0	47.5	52.5	55.0	52.5
Number of juveniles missing[c]	310	493	358	541	626	2,328
Estimated number of yearlings	111	177	128	194	225	835
Estimated number of adults[d]	81	129	93	141	164	607

a I assumed each female laid an average clutch of four eggs.
b The number of eggs and chicks lost is the difference between the number of eggs laid and the number of fledged chicks.
c I assumed a constant of 26.4% of all fledglings survived to become yearlings (Butler 1995).
d I assumed a constant of 72.7% of all yearlings survived to become adults (Butler 1995).

can be made, but they involve resighting or recapturing marked individuals (Clobert and Lebreton 1991). Herons are nearly impossible to catch, and their young are in nests difficult to reach by banders. I needed another method.

Juvenile and adult herons are easily distinguished in the field. By making regular censuses, an estimate of the local survival of these age classes can be derived. This method has its own biases – it assumes that all age classes have an equal chance of being sighted, that individuals that have left the survey route have died, and that the route chosen is representative of habitats used by all age classes. I assumed that age classes have an equal chance of being seen since herons are large birds and since my survey route covered representative habitat, but I did not know if birds not seen along the route had died. However, during periodic trips through farmland, I did not see herons move into other habitats in the delta after November.

Juvenile herons disappear more suddenly from censuses on the Fraser River delta throughout the autumn and winter than older herons. Nearly three-quarters of the juveniles counted in August had vanished by February, whereas only about one-quarter of the adults were missing (Butler 1995). Whether they died or dispersed to other locations was unknown. However, studies of other species of wading birds suggested that mortality among dispersing herons is great in the first few weeks after fledging. Mike Erwin and his colleagues (Erwin et al. 1996) attached small radio transmitters with mortality sensors to the backs of young egrets and night-herons in colonies in Virginia. Within fifty-five days of leaving the colony, 40 to 75 per cent of the chicks were dead. High mortality also occurs among recently fledged great blue herons. Powell and Bjork (1990, cited in Erwin et al. 1996) found that 90 per cent of fledglings in southern Florida died in their first six months of life. In summary, about half of all eggs laid became fledglings, about 14 per cent became yearlings, and about one in ten became an adult. The current information points to critical life and death decisions for a heron soon after it leaves the nest.

Causes of Mortality

Mortality among heron chicks begins at an early age. Veterinarian Ken Langelier performed postmortems on forty-three nestlings discovered below nests and found some surprising results (Butler 1995). Twenty-three chicks had died from a trauma, most likely the fall from the nest, as expected, but nine chicks had broken wings or legs that had partly healed before the fall from the nest. Another striking finding was that nearly half of the forty-three chicks' stomachs contained food. These chicks had not died from starvation. Only fifteen chicks were judged to be thin, of which six had food in the gizzard and only four had likely died from starvation. These results supported Doug Mock's (1985) contention that siblings kill one another for reasons other than food. However, food shortages might have become critical later if the chicks had survived.

Once herons leave the nest, mortality continues apace. About two-fifths of herons turned into wildlife rescue facilities in Vancouver had fractured bones, and about one-quarter were emaciated. Some had lacerations to the head, wings, or legs, some had been caught in fishing gear, and a small number had been shot or suffered from trauma (Butler 1995). Thus, great blue herons continue to be persecuted even though they have been protected by legislation since 1916. Some had a crippling disease to their feet known as 'bumblefoot,' while others had been poisoned or caught by dogs, and still others had burns to the head, bone dislocations, or a fish stuck in the gullet.

Few herons died directly from starvation. Ken Langelier's postmortems of twenty-nine juveniles showed that only four herons had likely starved to death and that the rest had either starved following an accident or died of accidental causes. In another sample of dead herons from the Fraser River delta, I considered a bird to have starved if it had less than one gram of fat in its visceral cavity. Thirty (64 per cent) of forty-seven juveniles were considered to have starved versus three (23 per cent) of thirteen adults. The most emaciated juveniles were 33 per cent below their estimated lean, fat-free body mass, whereas the fattest individual was 42 per cent above its lean mass. A starving heron was a sad sight. The flight muscles on the breast were severely depleted so that the sternum projected sharply from the breast. Some were unable to fly, and those that could get airborne were ungainly. The average liver of starving herons weighed about one-third less than livers from fat herons (Butler 1995). There was no statistically significant difference in the size of sixteen herons (wing, bill, and tarsus length) that died before a sudden spell of freezing weather and six herons found dead during and after the freeze.

Other studies of birds have shown that the first few weeks after departing from the nest are critical to their survival. Avoiding both starvation and predators is a problem for many birds, including the heron. The major predator of the heron in British Columbia is the bald eagle. Thousands of eagles that nest along the British Columbia coast mostly scavenge along beaches, but occasionally they hunt birds, including juvenile, yearling, and adult herons. My notebook contains a record of an incident at Sidney Lagoon on 15 June 1987:

> An adult Bald Eagle swept over juvenile great blue heron foraging in the lagoon. Heron took flight but the eagle drove it to the water, swooped four times, then immature Bald Eagle joined in, swooping twice at heron. Adult eagle chased immature eagle but a second joined in. Second immature swooped three times at heron and departed. Each time great blue heron flattened itself on water (ca. 50 cm deep). Ten herons soared and croaked from above. At 1554 h, juvenile took flight, circled the lagoon and returned to the colony.

On 18 and 19 July 1988, I observed an eagle kill two juvenile herons in the lagoon:

> Two juveniles killed by Bald Eagle adult (one today and one on 18 July). First juvenile was caught in mid-air by eagle and dragged to Eagle Island where it was eaten. Second juvenile was caught by eagle (in the air) and both fell into lagoon. Eagle flew to tree and heron remained for 5-10 minutes in water and then flew toward shore. Eagle pursued and both fell into water where eagle left the heron. Heron waded into forest and died.

Juvenile herons are very naïve about attacks by eagles and are thus easy prey, whereas adults recognize the threat of an approaching eagle intent on attacking ducks or other herons in the lagoon and take flight while the eagle is still far away. Other researchers have reported eagles killing herons in British Columbia (Butler 1989, 1995; Kelsall and Simpson 1979; Norman et al. 1989), and heron remains have been found at eyries in the Gulf Islands (Vermeer et al. 1989). Many heron colonies have been abandoned following eagle attacks (Butler 1995). Ian Moul witnessed an eagle on a freshly killed adult heron in the Sidney Island colony in 1990, and after repeated attacks the herons abandoned the island as a nesting site.

Population Growth and Limitation

Long-term data sets of the number of grey herons nesting in the Thames River basin in southern England have become a classic example used to describe how animal populations respond to changes in abundance. David Lack (1954) was the first to show that the number of nesting pairs crashed following harsh winter weather and rebounded to former levels following mild winters. Lack proposed that populations increase when food becomes more available to the survivors, which subsequently raise more offspring until the population reaches a point where food is in short supply. Subsequently, the number of adults that can breed in a colony and the number of young raised each year are reduced, and the population stabilizes. According to Lack, population growth of the Thames herons was checked by their ability to find food.

The census data now span many decades, and the relationships still hold – heron numbers fall after harsh winters and quickly rebound to former levels after mild winters (Perrins and Birkhead 1983). However, evidence that population growth is curtailed by food shortages has not been presented for herons along the Thames or anywhere else. This is not to say that the relationship proposed by Lack is spurious – only that convincing evidence is lacking. Evidence from grey herons in France suggests that populations there were limited by the number of territories rather than food abundance per se. Loic Marion (1989) showed that grey herons defended foraging sites against other herons, but he did not have the critical evidence that some herons were prevented from breeding as a result. Nevertheless, none of these results applies to herons in British Columbia, which are largely non-territorial; moreover, many previously used breeding sites are vacant (Butler 1995). And seashore habitats probably buffer any effect of harsh weather on adult herons in

British Columbia, whereas grey herons in freshwater habitats along the Thames River have no option but to await a thaw. Mechanisms that limit population growth likely differ between sites, so that no single explanation applies to all populations.

Earlier in this chapter, I mentioned that young egrets and herons suffer high rates of mortality soon after leaving their nests. Young herons are largely left on their own away from the nest – they are not fed on the beach by their parents or defended from approaching danger. The family bonds are quickly dissolved, and the young begin to roam in search of food. To survive, a young heron quickly has to learn many skills, including avoiding eagles and where and how to find food. Herons are very inept foragers in their first year. Soon after leaving the nest, they stand at the water's edge jabbing at bits of seaweed and twigs. Before long, they venture into deeper water and soon discover fish. Their clumsy strikes often come up empty. Overall, juvenile herons make more errors than adults when it comes to capturing prey. As the season progresses, juvenile herons become more adept at catching prey, but they still lag behind adults by midwinter. I suspect that this inability to catch food is the major cause of high mortality among young herons. Although they might not starve to death, their poor skills put them in dangerous situations. Robin Gutsell (1995) also showed that young herons have little choice but to forage on small mammals in grasslands during winter, for their poor fishing skills would ultimately result in their starvation. There they encounter electrical wires, fences, cars, and dogs. Poor foraging skills, naïve behaviour, and dangerous habitats are a bad mix for a heron.

Learning takes time and practice. A heron that fledges in June has a better chance of learning how to catch fish than a heron leaving a nest in August. In June, tides are among the lowest of the year for much of the day, and fish are abundant, but a heron that fledges in August has poor tides and fewer fish to catch. Sculpins and flounders make up a large part of the diet of herons, and by late summer some of these fish have grown large. The race for young herons is to keep foraging skills apace with declining opportunities to catch fish – the curtain of death drops suddenly for herons with lagging skills. Many young herons die long before food is in short supply to adults. This hypothesis assumes that the number of herons surviving their first year in British Columbia is closely associated with the rate at which they learn to catch fish on the beaches and not with the number of herons that seek the same food or the number of territorial herons (Butler 1994a).

For an adult heron, nesting early seems vital to success. Not only are early-nesting pairs more successful at raising young herons to fledging age, but their early-fledged young also arrive on beaches when food is more abundant and when tides allow long foraging periods. So what prevents adults from nesting earlier? Herons have reduced the courtship period to a matter of a few days. Elaborate displays and plumes that accentuate their abilities might have evolved to allow herons to make quick choices and get on with reproducing. Little time can be gained there. The key factors are a female's ability to catch food and then to produce eggs (Butler 1993). Age probably plays an important role here. Grey herons three or more years older raise more chicks than pairs in which one or both parents are younger than three years of age (Lekuona and Campos 1995). I would wager that older grey herons in this study nested earlier than younger herons.

Great blue herons must locate good foraging sites, choose a capable mate, locate a colony and nest site, find sufficient food, and avoid predators to survive and breed. This is a tall order! Great blue herons have evolved over thousands of years through natural selection to adapt to the vagaries of nature on the British Columbia seashore. In recent decades, a new challenge to the heron's adaptability has appeared on the coast – millions of humans who destroy and disrupt heron nesting and foraging habitats and pollute their food supplies. Can the heron cope with these changes? This is the subject of the next chapter.

CHAPTER 8

CONSERVATION OF HERONS AND THE STRAIT OF GEORGIA ECOSYSTEM

Before Europeans began to meddle in the Strait of Georgia, the hills and mountainsides were forested with Douglas-firs over ten centuries old, and wildflower meadows were strewn beneath garry oaks along the east coast of Vancouver Island. The Fraser River meandered through grasslands, brackish marshes, and bogs rimmed by ridges of crabapple, birch, pine, and spruce. R.T. Williams, writing in the *BC Directory* of 1884, quoted A.C. Anderson: 'Wild geese and ducks abound along the sloughs,' and 'wildfowl gather in vast numbers in late fall; Canada, white and crow goose, mallard, pintail.' Sandhill cranes bred in bogs and marshes in many places in the lower Fraser River valley. Brant were so plentiful during winter in eelgrass meadows in the northern Strait of Georgia earlier this century that Jim Spilsbury recalled, 'Many times I have seen the entire flock [of Brant] rise as one bird and literally darken the whole western sky. On a quiet night you could poke your head outdoors and hear the busy chattering of thousands and thousands of brant out on the reefs' (White and Spilsbury 1987). The whaling industry extirpated the humpback whale from the waters of the strait about the same time as the brant disappeared (Ketchen, Bourne, and Butler 1983). It is difficult to imagine the abundance of life that must have surrounded Aboriginal people. To the early Europeans, this abundance seemed limitless, and they harvested the bounty at an increasing rate until the majestic forests, whales, brant, sandhill cranes, and many fish stocks were all but gone.

By 1987, less than 1 per cent of the Fraser River delta was protected primarily for its wildlife (Butler and Campbell 1987), and a similar fate befell estuaries on Vancouver Island. However, renewed enthusiasm to preserve the remnant portions of natural and near-natural habitat gained momentum. In the 1970s and 1980s, estuarine marshes in the Little Qualicum, Nanaimo, Cowichan, and Campbell Rivers, and eelgrass meadows at Sidney Island, were secured in parks and conservation areas. A few years ago, eelgrass meadows stretching seventeen kilometres between Parksville and Qualicum were secured in a Provincial Wildlife Management Area. Several small islands near the mouth of the Fraser River and the immensely important beaches and eelgrass meadows for wildlife in Boundary Bay have been protected by legislation in the past five years. Roberts Bank and Sturgeon Bank in the mouth of the Fraser River are being considered by the BC cabinet as Provincial Wildlife Management Areas. Together with Boundary Bay, these Wildlife Management Areas are also being considered for designation as a Hemisphere Shorebird Reserve as part of an international venture to save a chain of important habitats known as the Western Hemisphere Shorebird Reserve Network. When completed, the protection of this complex of over 27,000 hectares of eelgrass meadow, mudflat, saltmarsh, and estuarine marsh will be the most significant conservation effort to benefit herons in British Columbia to date. As valuable as these actions have been, will they be sufficient to save the heron?

Threats

The survival of the great blue heron on the south coast of British Columbia is potentially threatened by the destruction of forests where they nest, by declining breeding success as a consequence of human disturbance at colony sites, and by the degradation of habitats that provide herons with food. There are about 4,200 adult herons on the coast of British Columbia, and about three-quarters of them nest on privately owned lands. Eight out of ten herons on the coast live around the Strait of Georgia, where the threat of disturbance and destruction of colonies and foraging habitats is great.

As a result, the great blue heron in British Columbia is considered 'at risk' by the Ministry of Environment, Lands and Parks. Species at risk are assigned a colour code of blue, so that blue-listed species 'are considered to be "vulnerable and at risk" but not yet endangered or threatened. Populations of these species might not be in decline, but their habitat or other requirements are such that they are vulnerable to further disturbance' (Harcombe et al. 1994). Species on the red list are endangered or threatened or are being considered for such status under the BC Wildlife Act. 'Any indigenous taxon

Many important habitats in the Strait of Georgia have been protected for herons and other birds. The marshes and grasslands within the Alaksen National Wildlife Area and George C. Reifel Migratory Bird Sanctuary at the mouth of the Fraser River hold large numbers of herons. These sites are maintained by Environment Canada's Canadian Wildlife Service and the British Columbia Waterfowl Society. *(Photograph by Rick McKelvey)*

(species or subspecies) threatened with imminent extinction or extirpation throughout all or a significant portion of its range in British Columbia is Endangered. Threatened taxa are those indigenous species or subspecies that are likely to become endangered in British Columbia if factors affecting their vulnerability are not reversed' (Harcombe et al. 1994). The Committee on the Status of Endangered Wildlife in Canada (COSEWIC) assigns a status of nationally vulnerable, threatened, or endangered to species and subspecies at risk in Canada. The status of the coastal subspecies of the great blue heron is 'vulnerable.'

The future for the heron does not look very rosy given the projected growth in the human population around the Strait of Georgia and Puget Sound and the rate at which nesting sites are being lost. The human population grew 9.1 per cent in the lower Fraser Valley between 1981 and 1986, or about 23,000 people per year (British Columbia Round Table on the Environment and the Economy 1993; Moore 1990). The consequence of this growth for heron habitats was clear – for every 1,000 increase in human population, twenty-eight hectares of rural land were converted into urban use (Moore 1990). The rate of human population growth on Vancouver Island and the Gulf Islands has been slightly lower than in the lower Fraser River valley. Soon there will be very few quiet woodlots on the south coast where herons can nest in peace.

Compounding the problem of human disturbance is the potential increase in disturbance from bald eagles. Eagles kill chicks and adult herons in colonies and juveniles at foraging grounds, and they disturb nesting adults, allowing corvids to take eggs. In 1987, Kees Vermeer, Ken Morgan, and I found over 100 nests and tallied over 700 immature eagles in the Gulf Islands (Vermeer et al. 1989). Using historical censuses, we estimated that the eagle breeding population had increased 30 per cent in the previous decade. And the eagle population continues to grow. There has been improved survival of younger age classes, which prey on the expanding gull population and scavenge in garbage dumps. The consequence of more eagles is increased disturbance at heron colonies.

As if disturbance from humans and eagles is not enough, herons in the Strait of Georgia now carry contaminants in their tissues (Elliott et al. 1996; Whitehead 1989). Most of the contaminants in heron eggs and tissues can be traced to the expansion of agriculture, manufacturing and forest industries, and urban developments. Some of these chemicals have been linked to

Urban sprawl has consumed large tracts of land on the Fraser River delta so that today housing (foreground) has edged into farmland bordering Boundary Bay (background). Housing and industrial development associated with growth in the human population threatens to destroy remnant woodlots where herons nest around the Strait of Georgia. *(Photograph by Dave Smith)*

reduced growth in heron chicks (Elliott et al. 1996). At present, we are unable to detect a strong negative effect in the population of herons around the Strait of Georgia.

Early Warning or False Alarm?

What evidence do we have that herons on the coast of British Columbia might be at risk? Some large colonies have grown in number in recent years, but a significant 6 per cent annual decline in the number of great blue herons was detected in data from Breeding Bird Surveys conducted on the Pacific coast between 1966 and 1994 (Downes and Collins 1996). Are these findings indicative of an early warning or a false alarm?

Let's start with growth in the large colonies. All the growth in the number of herons has occurred in two colonies in the Fraser River delta. Between 1977 and 1995, the respective annual average increase was 3.2 breeding pairs per year at UBC ($p = 0.008$) and 11.4 breeding pairs per year at Point Roberts ($p = 0.007$; see Figure 16). At the same time, the Stanley Park colony did not increase significantly in size. The UBC colony increased rapidly in 1977 and again in 1996, whereas the large increases in the Point Roberts colony occurred in 1990 and 1992 (see Figure 16). Thus, the annual changes in the number of breeding pairs of herons at colonies are not reflective of a series of good years; instead, they appear to be mostly colony specific.

Why did the breeding population of herons increase at UBC and Point Roberts but not at Stanley Park, and why did those increases occur in different years? The Point Roberts colony had two rapid increases of ninety-four pairs in 1990 and the eighty-seven pairs in 1992. These increases might be attributable in part to a few pairs that abandoned other colonies in the surrounding area, but most of the increase was probably due to many newly recruited first-time breeders. In the period 1977-95, about fourteen pairs of herons abandoned three colony sites on Sea Island and in north Delta, and other small colonies, unknown to me, might have been abandoned at the same time. These herons might have moved to Point Roberts, but it is unlikely that nearly 100 herons would have been overlooked. Can we account for the increases from what we know about the recruitment of first-time breeders? Most herons do not breed before two years of age, so all first-time breeding herons recruited at Point Roberts would have been born in 1988 and 1990. There were about 335 and 350 pairs of breeding herons respectively in the Point Roberts colony in those years. If we assume that each female laid an

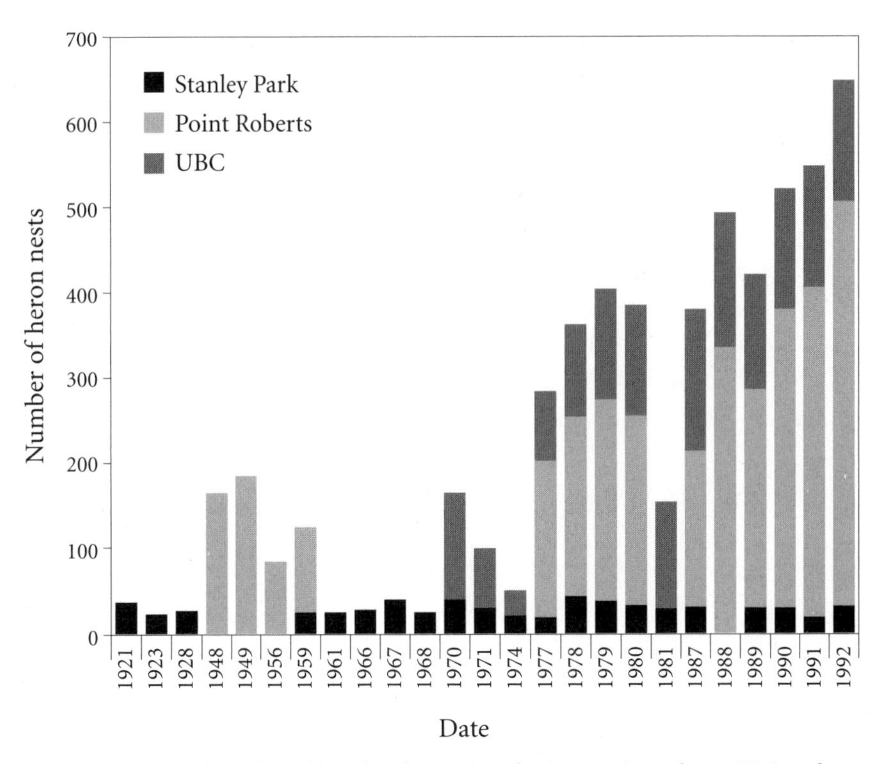

Figure 16 **Historical number of breeding herons in colonies at Point Roberts, UBC, and Stanley Park**

average clutch of four eggs and that one out of every ten eggs laid survived to become an adult, then the respective number of recruits of first-time breeders at Point Roberts in 1990 and 1992 would have been about 134 (0.4 multiplied by 335) and 140 (0.4 multiplied by 350) herons. Thus, recruitment of first-time breeders could explain the observed increases of breeding herons at Point Roberts in 1990 and 1992. This is good news for the large Point Roberts colony. However, there are many pairs that nest in colonies with much more variable reproductive success, and their survival might rely on recruits from large colonies such as at Point Roberts.

Most of the research on herons has been done at medium and large colonies, so our interpretation might be biased toward herons that nest in them. About half of all herons in British Columbia breed in colonies with long histories of use. Colonies at Stanley Park, Tillicum Mall, UBC, Nicomekl (includes Crescent Beach in Appendix 1), Holden Lake (includes Gabriola Island), Crofton, and Point Roberts or at alternative nearby sites have been used for twenty or more years. These colonies are generally characterized by large numbers of breeding pairs that are stable or growing in number and by pairs that reproduce successfully each year. All of these herons, except for those at Stanley Park, forage on large intertidal flats, where fish populations are predictably abundant each spring and summer. The Stanley Park colony is unusual in that its herons scatter along the seashore to forage.

A second group of herons nests in generally small colonies with fluctuating numbers of breeding pairs, short tenure of site use, and variable nesting success. About half of all herons in British Columbia nest in these colonies. Thus, within British Columbia about half of the herons exhibit a nomadic lifestyle, roaming between sites within and between breeding seasons, and about half nest at stable sites each year. Many small colonies use a site for only a few years, and the number of nesting pairs shows wide swings between years (Butler et al. 1995). Not all the former habitat is occupied, which suggests that the range of the heron has shrunk. Some formerly used foraging habitats at Pender Harbour (Kelsall and Simpson 1979; Simpson 1984) and Sidney Island have remained unoccupied or underutilized by herons for over fifteen and five years respectively.

Predicting the Impact of Disturbance

Given the information before us, can we begin to predict the impact of increased disturbance at heron colonies? Some herons are very sensitive to the presence of humans near their colonies. Pairs frightened from nests lose eggs to corvids, and late-nesting pairs fail to raise young herons more often than early-nesting pairs. If everything else is equal and no young are raised, the rate of population decline will be equal to the mortality rate of the breeding population, which I estimate to be about 25 per cent per year if the population is evenly aged. A breeding population of mostly young adults will decline more slowly than a population of older herons. Most of these demographic data are unknown for the coastal heron, but population declines of 25 per cent per year would most likely be detected during censuses.

With this information on disturbance of herons by eagles and humans, I posed the following problem to Christine Hitchcock, a colleague with an aptitude for computer modelling and a shared interest in herons (see Appendix 3). Imagine that 1,100 female herons nest in large colonies and that 970 females nest in small colonies, a situation similar to that reported in recent censuses. Each female raises 1.7 fledglings per year when the colonies are undisturbed and no fledglings in years with disturbance. Assume that 30 per cent of all surviving first-time breeders born in large colonies are recruited into small colonies, where they breed for the rest of their lives. Also assume that mortality rates remain unchanged. At present, about 5 per cent of herons in small colonies fail to raise young each year. In the model, Chris increased the disturbance effect so that 25 per cent of herons in small colonies fail to raise young each year. The computer simulation revealed a population that declines by about 10 per cent in a decade and 50 per cent in forty-five years (see Figure 17).

How much faith should we put in these simulations? After all, many of the assumptions are at best educated guesses. For all their shortcomings, though, the simulations suggest that the heron population will decline with increased disturbance. If this conclusion is correct, then the heron might become a red-listed species. Costs associated with red-listed species recovery can be immense, especially when recovery requires habitats in the highly

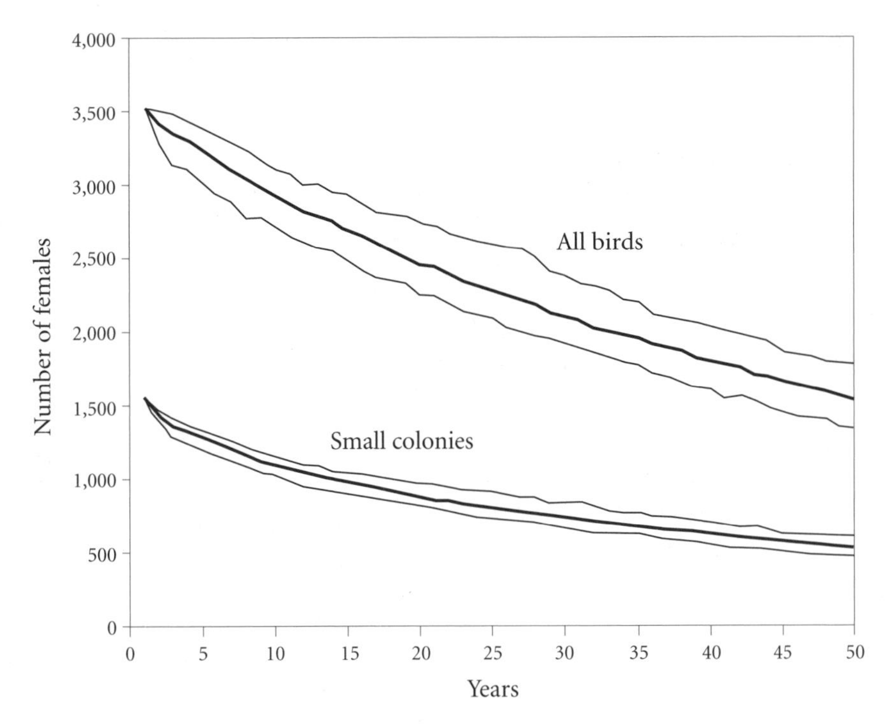

Figure 17 **Estimated number of breeding pairs of herons in the Strait of Georgia, with a 25 per cent increase in disturbance to small colonies**

Some herons have habituated to the presence of humans. This heron learned to beg handouts from zoo keepers at the former penguin pool in Stanley Park. However, many herons will leave their nests unguarded in the presence of humans. *(Photograph by Rob Butler)*

priced real estate market of southwestern British Columbia. Present levels of disturbance are unlikely to be alleviated for herons, so it is reasonable that conservation action should begin now while we have time and the costs are relatively low. In the meantime, we should refine our estimates used in the computer models of heron populations. Regular censuses of heron colonies – with the goal of estimating the proportion of the population that fails to reproduce – should be a priority. Estimates of survival of age classes are also important since changes in this parameter will alter population levels. We also need to know the level of migration between colonies and the quantity of nesting habitat remaining around the Strait of Georgia – this information needs to be included in a landscape-level plan for the conservation of the living resources of the Strait of Georgia. Along with this research, we need an on-the-ground conservation program.

Conservation Needs

The specific conservation needs of herons in British Columbia must include the preservation of quality foraging habitats and nesting sites. The major foraging habitats for adult herons in the Strait of Georgia are eelgrass meadows, estuarine and riverine marshes, and abandoned grasslands on the Fraser River delta. Eelgrass meadows support most of the coastal breeding herons in the province. Marshes are used year-round, and grasslands support many of the adult and juvenile herons in autumn and winter. Added bonuses in heron conservation programs are the many other important species associated with heron habitats, including internationally significant numbers of waterfowl, shorebirds, and birds of prey. Over 140 species of birds share the aquatic and terrestrial habitats frequented by herons.

Nesting sites for herons are ideally located in mature woodlots free from human disturbance and within three kilometres of foraging sites. The key feature of nesting sites is the distance above the ground rather than the species of tree – few herons build nests less than fifteen metres above the ground. Therefore, the tree community is probably less important to nesting herons than the age of the forest stand. Large spruces and firs used by herons for nesting sites are often centuries old, and the cottonwoods and alders are more than thirty years old. Some studies have shown that over time trees used by nesting herons die from their excrement, which inhibits photosynthesis and transpiration of the leaves, rather than from changes in soil pH and chemistry (Julin 1986). On our coast, the death of trees from heron excrement does not seem to be a problem, probably because it is washed away with the frequent rain.

Reducing human disturbance at heron colonies is an important consideration in a conservation plan for herons because of the negative effects of disturbance on nesting success (Vos, Ryder, and Graul 1985; Werschkul, McMahon, and Leitschuh 1976). Many authors have recommended forested buffer zones to exclude human activity around colonies while herons are nesting. These zones can range from as many as 1,000 metres (Koonz and Rakowski 1985) to as few as 250 metres (Vos, Ryder, and Graul 1985). However, forested buffer zones of this size might be unnecessary if other means of minimizing human disturbance are available. Becky Carlson and Bruce McLean (1996) showed that barriers that reduced human foot traffic under colonies had a stronger effect on the number of herons that fledged than buffer zones around colonies. In their study, barriers included fences, water, and land that made access difficult for humans. Heron colonies isolated by fences and water produced significantly more fledglings per nest than colonies with land buffer zones and colonies without buffer zones. In fact, the size of the buffer zone had no significant effect on the fledging success of herons. Carlson and McLean estimated the fledging success of nineteen heronries in Ohio and Pennsylvania with varying degrees of disturbance by humans. Each site was ranked as having no disturbance; mechanical disturbances such as farm equipment, vehicle traffic, trains, or low-flying aircraft; and foot traffic from hikers or horseback riders. The number of fledglings raised in heron colonies with frequent foot traffic was significantly lower than at colonies with no disturbance or mechanical disturbances. These results strongly suggest that some herons are very sensitive to the presence of humans in their colonies. Some authors have found that individual herons habituate to non-threatening human activities over time (Anderson 1978; Vos, Ryder, and Graul 1985), and some herons might be better able than others to cope with humans. For example, herons nesting in Stanley Park are unperturbed by throngs of people on the ground below. A few of these herons have even taken handouts from zookeepers, and one individual regularly followed Vancouver Aquarium staff into an enclosed walkway leading to a food storage area for marine mammals. In whatever method of conservation used, the important point is to keep disruptive human activities away from colonies during the nesting period between February and August. When implementing these guidelines, common sense should prevail. When there are doubts, a biologist experienced in dealing with herons should be consulted to recommend specific timing and proximity of activities around heron colonies.

The Heron Stewardship Program

The Heron Stewardship Program was launched in early 1997 by the Wild Bird Trust of British Columbia, with support from the Canadian Wildlife Service, BC Wildlife Branch, Habitat Conservation Trust Fund, and royalties from this book. The aim of the project is to secure sufficient habitat for the survival of the great blue heron around the Strait of Georgia through landowner participation. Public and private landowners with herons on their land are encouraged to join the Heron Stewardship Program by participating in one or more stewardship options, ranging from practical enhancement techniques to long-term agreements. In cases where a threatened site with a long history of use by herons is for sale, the Wild Bird Trust will consider land purchase. A team of volunteer wardens maintains and monitors sites in the Heron Stewardship Program. A similar program has been proposed by industry and government to protect herons in the Puget Sound region, thereby encapsulating the most important portions of the range of this subspecies.

Much has already been done in recent years to protect habitats used by herons, but the Heron Stewardship Program will address the important step of preserving heron habitats not included in other conservation programs. For example, many important heron habitats have been secured on the Fraser River delta, including Boundary Bay, Sea Island Conservation Area, Robertson Farm on Westham Island, and Spetifore Lands at Tsawwassen, Iona, Don, Lion Islands, and the South Arm Marshes. Sturgeon Bank and Roberts Bank at the mouth of the Fraser River will likely be added to this list of sites in the near future. They join several wildlife lands, such as the Alaksen National Wildlife Area in Delta and the Serpentine Fen in Surrey, that are used by herons. The Delta Farmland and Wildlife Trust has taken an important step in setting aside grasslands on farmlands that will benefit herons.

However, several nesting sites in the delta and around the Strait of Georgia have not been formally protected for herons. A low level of protection is afforded to a large colony at CFB Chilliwack by the Department of National Defence and to colonies at UBC, Stanley Park, Beacon Hill, and Tillicum from park status. However, the Point Roberts colony and most of the small colonies in the Fraser River delta and on Vancouver Island are on privately owned land, and protection is left largely to the whim of the owners.

The Heron Stewardship Program will provide increased protection to lands of importance to herons while benefiting many internationally important wildlife populations. Although this program and others with similar conservation goals buy important time, the swell of human population threatens to undermine these good intentions. It is unlikely that healthy wildlife populations can be maintained in perpetuity if human consumption is not reduced (Mangel et al. 1996). The 'museum' concept of saving living resources in reserves will not work for many species. The size of the area required is just too large to be accommodated in some parts of the world. One of the mechanisms for conservation proposed by Mangel et al. (1996) is to identify areas, species, and processes that are particularly important to the maintenance of ecosystems; minimize fragmentation; and avoid disruption of food webs, especially the removal of top species. Attempts to reduce the industrial, agricultural, and urban discharges into the Fraser River have had some success, but the increasing quantity of effluent flowing from the rapidly growing human population in the region threatens to overtake these improvements (Fraser Basin Management Program 1996). The greatest fear is that pollution might uncouple the ecological functions of important wildlife habitats downstream and undermine the habitats that many people have struggled so hard to preserve.

Few other animals better symbolize a vision of conservation for the Strait of Georgia ecosystem than the great blue heron (see Plate 9). It lives year-round on the shores of the strait, wades on its beaches and in its streams, rivers, and marshes, hunts in grasslands and from kelp forests, nests in old-growth rain forests, and penetrates the urban landscape. As sentinels, the heron's eggs provide a means to monitor contaminants in the rivers and ocean, and its reproductive success might just provide clues to the abundance of fish in inshore waters. Conserving the heron and its environment would go a long way toward ensuring the conservation of much of the quality of life in the Strait of Georgia and Puget Sound.

EPILOGUE

I BEGAN THIS BOOK by correcting Bruce Hutchison's view that the Fraser River had produced no myths, no songs, and no rivermen. Hutchison had overlooked the Stō:lo people. It seems somewhat ironic that fifty years later it is Aboriginal artists who lend us their symbols to carry proudly into orbit. Before Canadian astronaut Robert Thirsk traveled into space aboard the space shuttle *Columbia* in 1996, he spent considerable time selecting items representative of the spirit of Canada to take along on the voyage. Thirsk was clear in his intention: among the many symbols of Canada available to him, he selected a small brooch depicting an eagle totem, a bracelet, and a pendant depicting the moon, by Tsimshian artist Bill Hilen. Thirsk's choice trumpeted two messages: that West Coast culture had matured beyond European definition, and that animal images are a worthy depiction of us as a people. These messages would not have been heard even twenty years ago. Such a shift in cultural definition and in the desire to use animal symbols to depict us would not surprise anthropologist Wade Davis (1992). In his view, cultures are deeply rooted in their environments, and given enough time people adopt lifestyles from their surroundings. 'Just as landscape defines a people, culture springs from a spirit of place,' Davis wrote. Within Thirsk's choice of symbols is a third message that speaks to the need for conservation. Although Aboriginal people have been saying it for years, many non-Aboriginals are just beginning to realize that the environment provides more than just resources. If this hypothesis is correct, then the bad news is that degradation of our environment has diminished us as a people. The good news is that restoring the environment – so that majestic creatures such as herons continue to flourish – will repay us in dividends that we are only beginning to comprehend.

Each spring and summer, hundreds of herons fly between feeding grounds on Boundary Bay and their colony at Point Roberts. Paul Boeth was a quiet man liked by all who knew him. And he had a passion for herons. I did not have the privilege of meeting him, and on 25 June 1991, Paul Boeth died at his home in Tsawwassen. However, his passion lives on in his poetry.

Garden Flight

In my garden,
My eyes are downcast,
Whether tilling the soil
Or admiring the beauty
That my effort has produced,
My eyes are earthbound.
But, like a scene from some mythology book,
A great shadow of a bird
Glides past me toward the west,
And I find myself forced to look skyward.
A nest-bound great blue heron floats
Effortlessly by;
My thoughts fly with it.
After Man has driven this great bird away,
Will I be complacent about its absence?
Will I chalk it up to progress?
When development has
Homogenized its nesting area,
What great shadow will lift my eyes to heaven?
How will I send my thoughts flying,
With no great birds to use for a guide?

– Paul Boeth

10 (previous page) On the Pacific Coast of North America, a distinct subspecies of great blue heron, *Ardea herodias fannini*, has evolved that is characterized by being smaller and generally darker on the neck and upper parts than other great blue herons. *(Photograph by Wayne Campbell)*

11 (above) The great blue heron is adapted to hunting small fish in shallow water. Its long legs allow it to pursue fish easily; its eyes can shift focus to prey in front or below without moving the head; its long neck can be unleashed to strike forward; and its bill works like finely tuned tweezers to snatch wriggling fish from the water. *(Photograph by Tim Fitzharris)*

12 (facing page) The great blue heron catches small fish by waiting or slowly stalking them in shallow water. The neck is often outstretched while searching for prey. (A) Once a fish is detected, the heron slowly coils its neck as it steps closer to the prey. (B) The heron strikes forward and down with its head into the water, and (C) on about two-thirds of all attempts, it captures a fish. *(Photographs by Tim Fitzharris)*

A

B

C

13-14 (below) The sixty heron species in the world evolved from a subtropical ancestor. All species share a similar design shape of relatively long legs and neck, long pointed bill, and large wings. An all-white form, known earlier as the great white heron, inhabits extreme southern Florida and parts of the Caribbean. *(Photographs by Tim Fitzharris)*

15 (facing page) The great blue heron has become adapted to year-round existence on the rugged British Columbia coast. It forages from floating kelp, wades on shallow beaches, stalks small mammals in grasslands, and nests in mature forests. *(Photograph by Tim Fitzharris)*

16 (top) Nestling herons require large amounts of food for growth. A heron chick will grow from 50 grams to over 2,000 grams in about a two-month period. The demand this growth puts on parents is daunting – at about one month of age, a pair of herons with chicks consume about the same amount of energy as the average North American adult. *(Photograph by Wayne Campbell)*

17 (bottom) In *The Herons Handbook* (1984), James Hancock and James Kushlan write, 'Herons are amongst the most glamorous of all birds, giving grace and line to the watery landscape.' The graceful appearance of this display is to threaten an interloper. *(Photograph by Tim Fitzharris)*

18-23 (above) Great blue herons in British Columbia eat many fish and other animals. During the breeding season, the principal fish of the herons' diet are (A) the shiner perch (*Cymatogaster agreggata*), (B and C) gunnels (*Pholis laeta* and *Pholis ornata*), (D) bay pipefish (*Signathus griseolineatus*), (E) three-spined stickleback (*Gasterosteus aculeatus*), and staghorn sculpin (*Leptocottus armatus*). In autumn and winter, herons augment their fish diet with meadow voles, especially the (F) Townsend's vole (*Microtus townsendii*). *(Photographs by Danny Kent [A-D], Ernest Cooper [E], and Mary Taitt [F])*

24 (overleaf) Young great blue herons begin to moult into the characteristic white crown feathers of yearlings and adults about six months after they leave the nest. This juvenile shows many characteristics of an immature heron, including the slate-grey-coloured feathers on the crown, rufous-coloured margins to feathers on the wing, and few feather plumes. Young herons are not very efficient at catching fish, as this photograph shows. *(Photograph by Tim Fitzharris)*

How will I send my thoughts flying, with no great birds to use for a guide?
– Paul Boeth. *(Photograph by Tim Fitzharris)*

APPENDIX I

Historical records of great blue heron colonies on the British Columbia coast

Colony name	Year	Active (+) or inactive (–)	Number of nests	Number of occupied nests	Number fledged per successive nests	Number fledged per nest attempt	*N*	Latitude	Longitude
Fraser Valley									
Boundary Bay	1996	unknown	14					49°05′	123°21′
Burnaby Lake	1933	+						49°15′	122°56′
Burnaby Lake	1942	+							
Burnaby Mountain	1992	+		c. 20				49°17′	123°00′
Burnaby Mountain	1996	+		c. 20					
CFB Chilliwack	1976	+	50+					49°06′	122°03′
CFB Chilliwack	1977	+	96	88	2.62		34		
CFB Chilliwack	1978	+	101						
CFB Chilliwack	1979	+	110	109					
CFB Chilliwack	1980	+	104	91	2.63		43		
CFB Chilliwack	1992	+	152	58	2.81		58		
CFB Chilliwack	1993	+	105	102					
CFB Chilliwack	1994	+	160	113					
Cheam Lake	1992	+	13					49°11′	121°46′
Cheam Lake	1993	+	10						
Cheam Lake	1994	+	10						
Chilliwack City	1993	+	5					49°11′	121°56′
Chilliwack City	1994	+	15	12					
Clayburn, Matsqui	1993	+	6+	2				49°04′	122°16′
Clayburn, Matsqui	1994	–							
Colony Farms	1992	+	4	4	4.0		4	49°4′	122°48′
Colony Farms	1993	+	1	1	0	0	1		
Colony Farms	1994	–	1	0					
Crescent Beach	1967	+						49°04′	122°51′

Crescent Beach	1968	+							
Crescent Beach	1977	+	39	37	2.94		18	49°04´	122°51´
Crescent Beach	1978	+	46		2.76		17		
Crescent Beach	1979	+		42	3.1		13		
Crescent Beach	1980	+	41		2.27		22		
Crescent Island	1962	+	3					49°07´	123°04´
Crescent Island	1963	+							
Crescent Island	c. 1960	–							
Crescent Island	1974	+	1						
Crescent Island	1990	+	6+	0	0	0	6+	49°10´	122°26´
Crescent Island	1991	–	6+	0					
Debouville Slough - Site 1	1992	+	c. 70	47	0		59	49°17´	122°43´
Debouville Slough - Site 1	1993	+	1	34	0	0	34		
Debouville Slough - Site 2	1992	+	c. 50	41	1.89				
Debouville Slough - Site 2	1993	+	1	34	0	0	34		
Debouville Slough - Site 2	1994	–	34	0					
Dewdney 1	1972	+		10-11				49°14´	122°38´
Dewdney 1	1973	+		9					
Dewdney 1	1974	–							
Dewdney 2	1963	+		20				49°15´	122°37´
Dewdney 2	1970	+		30					
Dewdney 2	1971	–							
Dewdney 3	1973	+		9				49°16´	122°35´
Dewdney 3	1974	–		0					
Douglas Island	1973	+						49°13´	122°47´
Douglas Island	1977	+							
Edgewater Bar	1974	+	17					49°12´	122°37´
Edgewater Bar	1975	+	18						
Edgewater Bar	1976	+	20						
Edgewater Bar	1977	+		16+					
Edgewater Bar	1978	+	31						
Edgewater Bar	1979	+	38	38	3.1		12		
Edgewater Bar	1980	+	38	30	2.4				
Edgewater Bar	1981	+	33	31	2.1		15		

Appendix 1 Historical records of great blue heron colonies on the British Columbia coast, continued

Colony name	Year	Active (+) or inactive (–)	Number of nests	Number of occupied nests	Number fledged per successive nests	Number fledged per nest attempt	*N*	Latitude	Longitude
Fort Langley	1944	+						49°10′	122°35′
Fort Langley	1963	+	10-12						
Fraser Mills	1960	+	45	38				49°14′	122°52′
Harrison River	1992	+	63	32	1.89		45	49°18′	121°50′
Harrison River	1993	+	40+					49°16′	121°57′
Harrison River	1994	+	40+						
Hope - Greenwood 1	1973	+						49°22′	121°27′
Hope - Greenwood 1	1979	+	10						
McCormick Marsh	1989	+	40					49°14′	122°40′
McCormick Marsh	1992	–	12						
McGillivray Slough	1920	+						49°08′	122°06′
McGillivray Slough	1974	unknown	27						
McGillivray Slough	1977	+	51	46					
McMillan Island	1964	+	31-5					49°11′	122°34′
Mary Hill Bypass	1992	+	11	11	1.45		11	49°17′	122°42′
Mary Hill Bypass	1993	+	22						
Mary Hill Bypass	1994	+	54						
Mount Lehman	1988	+						49°08′	122°23′
Mount Lehman	1970	+	10						
Nicomekl River	1988	+		38		2.1 (1.3)	30	49°05′	122°49′
Nicomekl River	1990	+		40		2.2 (1.1)	37		
Nicomekl River	1992	+	34	34		3.2 (1.1)	34		
Nicomekl River	1993	+	36						
Nicomekl River	1995	+	35+						
North Alouette River	1974	+	10					49°15′	122°35′
North Alouette River	1991	–	70+	0	0	0	70+		
North Alouette River	1992	–							
North Delta	1992	+	6	6	0		6	49°09′	122°56′

North Delta	1993	–							
North Vancouver	1923	+	18					49°19′	123°08′
North Vancouver	1929	+							
Pitt Meadows - Crane	1979	+	1	1	1			49°18′	122°41′
Pitt Meadows - Crane	1980	–							
Pitt Meadows - Crane	1981	–	5						
Pitt Meadows - McIvor	1977	+	10	10	1.88		8	49°15′	122°39′
Pitt Meadows - McIvor	1978	+	10	8	2.13		6		
Pitt Meadows - McIvor	1979	+	8	5	3		5		
Pitt Meadows - McIvor	1992	+	c. 10						
Pitt Meadows - McIvor	1993	+	30	25				49°14′	122°40′
Pitt Meadows - McIvor	1994	+	43	43					
Pitt Meadows - Reavie Road	1992	–	6						
Pitt Meadows - Reavie Road	1993	+	6	4	0	0	6		
Pitt Meadows - Rippington Road	1979	+						49°15′	122°39′
Pitt Meadows - Rippington Road	1980	+	6	4	3.25		4		
Pitt Meadows - Rippington Road	1981	+	8	6	2.5		4		
Point Roberts	1948	+	165					49°05′	123°03′
Point Roberts	1949	+	185						
Point Roberts	1955	+							
Point Roberts	1956	+	>85						
Point Roberts	1958	+							
Point Roberts	1959	+	100					49°05′	123°04′
Point Roberts	1960	–							
Point Roberts	1961	+	some						
Point Roberts	c. 1965								
Point Roberts	1977	+	216	183	2.79		39	48°59′	123°04′
Point Roberts	1978	+	226	210	2.51		39		
Point Roberts	1979	+	247	236	2.9		33		
Point Roberts	1980	+	250	222	2.94		53		
Point Roberts	1981	+		220+	2.43		50		

Appendix 1 Historical records of great blue heron colonies on the British Columbia coast, continued

Colony name	Year	Active (+) or inactive (–)	Number of nests	Number of occupied nests	Number fledged per successive nests	Number fledged per nest attempt	*N*	Latitude	Longitude
Point Roberts	1982	+							
Point Roberts	1983	+							
Point Roberts	1984	+							
Point Roberts	1985	+							
Point Roberts	1986	+	220						
Point Roberts	1987	+	253	183		2.3 (0.8)	30		
Point Roberts	1988	+	335			2.0 (0.9)	55		
Point Roberts	1989	+	256			1.9 (1.2)	46		
Point Roberts	1990	+	350			2.1 (1.0)	61		
Point Roberts	1991	+	387			2.2 (0.9)	60		
Point Roberts	1992	+	474		3.04		25		
Point Roberts	1993	+	400						
Point Roberts	1994	+	414						
Point Roberts	1995	+	472						
Poplar Island	1973	+						49°12′	122°56′
Poplar Island	1974	+	10						
Port Coquitlam IR 1	1973	+	59	48+				49°14′	122°48′
Port Coquitlam IR 1	1977	+	169	169	2.31		42		
Port Coquitlam IR 1	1978	+	162	162	2.3		43		
Port Coquitlam IR 2	1979	+	34	31	1.76		17	49°15′	122°48′
Port Coquitlam IR 2	1980	+		26+	2.12		26		
Port Coquitlam N.Park	1971	+	76	68	2.8		68	49°17′	122°48′
Port Coquitlam N.Park	1973	–							
Richmond No. 7 & C	1984	+	1					49°10′	123°15′
Richmond No. 7	1987-96	+	7						
Richmond No. 7	1989	+	16	12					
Richmond No. 7	1991	+		12	0	0	12		
Richmond No. 7	1992	+	12	6	0	0	6		

Sea Island 2	1987	+	7	7					
Sea Island 2	1992	?	12						
Serpentine River	1994	+	5						
Stanley Park	1921	+	37					49°18′	123°09′
Stanley Park	1923	+	23						
Stanley Park	1928	+	27						
Stanley Park	1959	+	25						
Stanley Park	1961	+	25+						
Stanley Park	1966	+	28						
Stanley Park	1967	+	40						
Stanley Park	1968	+	25						
Stanley Park	1969	+							
Stanley Park	1970	+	40						
Stanley Park	1971	+	30						
Stanley Park	1974	+	21						
Stanley Park	1977	+	31	19	2.33		12	49°18′	123°08′
Stanley Park	1978	+	44	44	2.55		22		
Stanley Park	1979	+	40	38	2.4				
Stanley Park	1980	+		33	2.05		19		
Stanley Park	1981	+	29	20+	2.18		9		
Stanley Park	1987	+		31		3.5 (3.3)	16		
Stanley Park	1989	+		30		1.9 (1.2)	20		
Stanley Park	1990	+		30		2.5 (1.0)	15		
Stanley Park	1991	+		19					
Stanley Park	1992	+		32		2.8 (1.0)	32		
Stanley Park	1994	+	17	16	2.4		16		
Stanley Park	1995	+		19	2.6		19		
Stave River - Ruskin	1925	+						49°11′	122°25′
Stave River - Ruskin	1963	+		20-30					
Stave River - Ruskin	1964	+		35					
UBC Marine Drive	1970	+	125 - 130					49°14′	123°13′
UBC Marine Drive	1971	+		67	2.6				
UBC Marine Drive	1972	+	30						
UBC Marine Drive	1973	+							

Appendix 1 Historical records of great blue heron colonies on the British Columbia coast, continued

Colony name	Year	Active (+) or inactive (–)	Number of nests	Number of occupied nests	Number fledged per successive nests	Number fledged per nest attempt	*N*	Latitude	Longitude
UBC Imperial Drive	1974	+						49°15′	123°12′
UBC Imperial Drive	1977	+	92	82	2.91		23		
UBC Imperial Drive	1978	+	128	108	2.77		30		
UBC Imperial Drive	1979	+	130	130	2.91		22	49°16′	123°12′
UBC Imperial Drive	1980	+	147	130	2.4		51		
UBC Imperial Drive	1981	+	150	125+	1.93		42		
UBC Imperial Drive	1987	+	214	158		1.7 (0.9)	117		
UBC Imperial Drive	1988	+		158		2.1 (1.3)	18		
UBC Imperial Drive	1989	+		135		1.5 (1.0)	37		
UBC Imperial Drive	1990	+		141		2.2 (0.9)	18		
UBC Imperial Drive	1992	+		142		2.0 (1.2)	49		
UBC Imperial Drive	1993	+							
UBC Imperial Drive	1994	+		169					
UBC Imperial Drive	1995	+							
UBC Imperial Drive	1996	+		217					
UBC Research Forest	1974	+	10					49°17′	122°35′
UBC Research Forest	1975	+	10						
UBC Research Forest	1977	+		10+					
UBC Research Forest	1978	+		10+					
UBC Research Forest	1979	+		10+				49°17′	122°35′
UBC Research Forest	1980	+	13+	10+					
White Rock	1962	+						49°01′	122°48′
Widgeon Creek	1979	+	2+					49°21′	122°38′
Vancouver Island									
Beacon Hill	1988	+		64		1.7 (.8)	49	48°25′	123°21′
Beacon Hill	1989	+		55		1.4 (1.3)	33		
Beacon Hill	1990	+		27	0	0	27		

Bowser	1975	+	18					49°27′	124°41′
Bowser	1976	–							
Chatham Island	1929	+						48°26′	123°15′
Chehalis River	1957-60	+						49°06′	121°57′
Chehalis River	1979	unknown	84					49°17′	121°57′
Cherry Point	1939	+						48°43′	123°33′
Cobble Hill	1978	+		5-7				48°42′	123°35′
Cobble Hill	1979	+		1					
Cortes Island	1979	+	22					50°02′	124°59′
Cortes Island	1980	+		0					
Courtenay	1980	+	19					49°41′	124°59′
Cowichan	1989	+		1	0	0	1	48°48′	124°00′
Cowichan	1990	+		1	0	0	1		
Cowichan Bay Village	1931	+	8+					48°44′	123°38′
Cowichan Bay Village	1945	+	2					48°45′	123°39′
Cowichan Bay Village	1952	+	32						
Cowichan Bay Village	1973	+		12					
Cowichan Bay Village	1979	+	34	28					
Cowichan Station	1972	+	9	7				48°44′	123°44′
Cowichan Station	1973	+		12					
Cowichan Station	1975	+		27					
Cowichan Station	1977	+	7-10						
Cowichan Station	1979	unknown							
Crofton	1955	+						48°51′	123°38′
Crofton	1970	+	40						
Crofton	1973	+	59	52					
Crofton	1974	+	130	85					
Crofton	1975	+	75+						
Crofton	1979	+							
Crofton	1990	+	32			1.9 (1.7)	22		
Crofton	1991	+	47			1.8 (1.3)	42		
Crofton	1992	+	61			0.7 (1.1)	52		
Denman Island N	1979	+						49°35′	124°50′
Denman Island N	1980	+	8-10						

Appendix 1 Historical records of great blue heron colonies on the British Columbia coast, continued

Colony name	Year	Active (+) or inactive (–)	Number of nests	Number of occupied nests	Number fledged per successive nests	Number fledged per nest attempt	*N*	Latitude	Longitude
Denman Island S	1979	+						49°35′	124°50′
Denman Island S	1980	+	40-50						
Denman Island S	1981	+	70						
Denman - Allen A	1989	+		5		0.8 (1.5)	4	49°35′	124°50′
Denman - Allen A	1990	+		1		0	1		
Denman - Allen B	1989	+		15		2.8 (1.0)	13		
Denman - Allen B	1990	+		18		0	18		
Denman - Allen C	1989	+		4		2.0 (1.4)	4		
Denman - Allen C	1990	+		3		1.0 (1.7)	3		
Denman - Allen C	1991	+		14		0.3 (0.7)	14		
Denman - Brewer	1989	+		24		2.1 (1.6)	18	49°32′	124°48′
Denman - Brewer	1990	+		25		1.9 (1.8)	24		
Denman - Brewer	1991	+		22		1.3 (1.4)	22		
Denman - Elliott	1989	+		15		1.0 (1.4)	11	49°32′	124°46′
Denman - Elliott	1990	+		18		0.4 (1.1)	15		
Fanny Bay	1988	+	15	15		0.5 (1.1)	13	49°31′	124°50′
Gabriola Island	1979	+	c. 30					49°09′	123°49′
Gabriola Island	1980	+	c. 60						
Gabriola Island	1981	+	90	64					
Genoa Bay	1970	+	16	13				48°46′	123°36′
Genoa Bay	1971	+	22	22					
Genoa Bay	1972	+	17	10					
Genoa Bay	1975	–							
Genoa Bay	1979	+	9					48°46′	123°36′
Gibsons - town site	c. 1977	+	10					49°24′	123°30′
Gibsons - town creeks	1978	+	54					49°30′	123°30′
Gibsons - town creeks	1979	+	a few						
Gibsons - town creeks	1980	+		2					

Hammond Bay	1988	+		24	1.0 (1.0)	21	49°14′	123°59′
Hammond Bay	1989	+		16	1.4 (1.2)	12		
Hammond Bay	1990	+		21	2.3 (1.4)	13		
Hammond Bay	1991	+		15	0	15		
Holden Lake	1988	+		100	1.8 (0.7)	44	49°06′	123°50′
Holden Lake	1989	+		162	2.1 (1.0)	45		
Holden Lake	1990	+		132	3.0 (0.6)	36		
Holden Lake	1991	+		61	1.4 (0.9)	37		
Holden Lake	1992	+		112	2.0 (0.8)	78		
Hornby Island	1959	+	30				49°32′	124°40′
Hornby Island	1960	+						
Hornby Island	1961	+	c. 30	30				
Hornby Island	1962	+		16				
Hornby Island	1966	+	34					
Hornby Island	1962	+	1	1				
Hornby Island - Lazare	1989	+		7	3.1 (0.7)	7		
Hornby Island - Lazare	1990	+		9	2.2 (1.3)	5		
Hornby Island - Lazare	1991	+		10	0	10		
Hornby Island - Paris	1989	+		15	2.6 (0.9)	9		
Hornby Island - Paris	1990	+		17	0.6 (1.2)	17		
Hornby - Wiseman	1989	+		18	2.8 (0.9)	18		
Hornby - Wiseman	1990	+		9	0.3 (1.0)	9		
Hornby - Wiseman	1991	+		26	1.7 (1.1)	25		
Ladysmith	1939	+					49°00′	123°50′
Little River	1989	+		22	2.9 (1.4)	17	49°44′	124°54′
Little River	1990	+		27	3.9 (0.6)	16		
Little River	1991	+		28	2.3 (0.8)	25		
Little River	1992	+		44	2.3 (0.7)	32		
Mandarte Island	1945	+	7				48°38′	123°17′
Mandarte Island	c. 1936	+						
Mandarte Island	c. 1950	+						
Nanaimo - Cedar Road	1981		60+				49°07′	123°51′
Nanaimo - Extension	1972	+	3-4				49°06′	123°58′
Nanaimo - Extension	1980	+					49°16′	124°11′

Appendix 1 Historical records of great blue heron colonies on the British Columbia coast, continued

Colony name	Year	Active (+) or inactive (–)	Number of nests	Number of occupied nests	Number fledged per successive nests	Number fledged per nest attempt	*N*	Latitude	Longitude
Nanaimo - Extension	1981	+							
Nanaimo - Yellowpoint	1972	+	10					49°03′	123°45′
Nanoose	1980	+	c. 8					49°16′	124°11′
North Pender Island	1974	+	1	1	1			48°47′	123°17′
Parksville	1981	+	5-7	5-7				49°19′	124°20′
Parksville - Acacia	1989	+		1		4.0 (0)	1	49°18′	124°13′
Parksville - Acacia	1990	+		1		4.0 (0)	1		
Parksville - Gravel Pit	1989	+		9		2.6 (1.8)	9	49°19′	124°16′
Parksville - Gravel Pit	1990	+		10		0	10		
Parksville - Northwest Bay	1989	+		8		2.4 (1.3)	8	49°12′	124°17′
Parksville - Northwest Bay	1990	+		11		3.5 (0.9)	8		
Portage Inlet	1977	+	5+					48°27′	123°26′
Port Alberni	1971	+	10					49°15′	124°49′
Prevost Island	1974	+	59					48°50′	123°22′
Quadra Island	1979	+	8					50°05′	123°15′
Quadra Island	1980	+							
Qualicum	c. 1900	+	large colony					49°20′	124°30′
Quamichan Lake	1989	+		32		2.2 (1.3)	27	48°49′	123°40′
Quamichan Lake	1990	+		28		2.8 (0.6)	20		
Quamichan Lake	1991	+		28		2.0 (1.0)	27		
Quamichan Lake	1992	+		19	0	0	19		
Sahtlam	1989	+	6			1.5 (1.2)	6		
Sahtlam	1990	+	3			1.0 (1.0)	3	48°48′	123°50′
Saltspring Island North								48°56′	123°34′
Saltspring Island Southey Point	1980	+	4-5					48°56′	123°35′
Saltspring Island Southey Point	1989	+	25			1.0 (1.4)	21		

Saltspring Island North Beach Road	1997	+	118					48°55′	123°34′
Saturna Island	1920	+						48°47′	123°08′
Secretary Island	1932	+	1					48°58′	123°36′
Secretary Island	1933	+	4						
Secretary Island	1934	+	12						
Secretary Island	1935	+	44						
Shoal Island	1982	+	20					48°54′	123°40′
Sidney Island	1974	+	30					48°37′	123°19′
Sidney Island	1979	+	75						
Sidney Island	1984	+	41		0	0			
Sidney Island	1986	+	86	86					
Sidney Island	1987	+		103		1.6 (1.1)	78		
Sidney Island	1988	+		99		1.8 (1.1)	75		
Sidney Island	1989	+		44	0	0	44		
Sidney Island	1990	+		16	0	0	16		
Sooke - Otter Point	1960	+	6					48°24′	123°45′
Sooke - Otter Point	1961	+	8	8					
Sooke - Stoney Creek	1961	+	2					48°24′	123°39′
Sooke - Whiffen Spit	1960	+	23	12				48°21′	123°43′
Sooke - Whiffen Spit	1961	+	20	12					
South Pender Island	1975	+	1	1	2		1	48°45′	123°12′
Swartz Bay	1963	+	5					48°41′	123°24′
Swartz Bay	1970	+							
Tahsis	1989	+		9				49°55′	126°9′
Thetis Lake	1962	+	8					48°29′	123°26′
Thetis Lake	1974	+						48°28′	123°28′
Tillicum	1983-5	+	20-60					48°27′	123°24′
Tillicum	1986	–	20-60	0	0				
Tillicum	1987-8	+	20-60						
Tillicum	1989	+		44		2.0 (1.5)	19		
Tillicum	1990	+		21		1.4 (1.6)	20		
Tillicum	1991	+		24	0	0	13		
Tillicum	1992	+		17		1.6 (0.8)	14		

Appendix 1 Historical records of great blue heron colonies on the British Columbia coast, continued

Colony name	Year	Active (+) or inactive (–)	Number of nests	Number of occupied nests	Number fledged per successive nests	Number fledged per nest attempt	*N*	Latitude	Longitude
Tillicum	1995	+	20-60	0	0				
Tsehum Harbour	1957	+						48°40′	123°25′
Union Bay	1988	+	6	6		1.0 (1.4)	4	49°35′	124°53′
Sunshine Coast and Howe Sound									
Anvil Island	1972-4	+						49°31′	123°18′
Anvil Island	1980	unknown	8						
Brackendale	1993	+	5					49°46′	123°10′
Gibsons, Stewart Road	1993	+	4	4	0	0	4	49°54′	124°36′
Gibsons, Stewart Road	1994	+	2	2	0	0	2		
Halfmoon Bay	1993	+	16	14	1.7	1.4	16	49°32′	123°54′
Lion's Bay	1980	2	4	4				49°25′	123°14′
Nelson Island	1980	–						49°43′	124°04′
Nelson Island	c. 1977	unknown							
Pender Harbour	1963-77	+						49°37′	124°00′
Pender Harbour	1978	+	93	43	2.1		19		
Pender Harbour	1979	+	94	44	3.0		18		
Pender Harbour	1980	–	–	–					
Pender Harbour	1981	+	25-35	0				49°39′	123°58′
Pender Harbour, Paq Lake	1993	+	10					49°37′	124°01′
Powell River	1973	+	30					49°52′	124°32′
Powell River	1974	+		14					
Powell River	1978	+	6+					49°46′	124°27′
Powell River	1980	+	19	19					
Powell River	1988	+	42	24	0	0	24		
Powell River	1990	+		38		1.6 (1.3)	33		
Powell River	1991	+		13		0	13		
Powell River, Glacier Street	1993	+	4					49°54′	124°31′

Powell River, Glacier Street	1994	+	7						
Powell River, Myrtle Rocks	1993	+	6					49°48′	124°29′
Powell River, Willingdon Beach	1993	+	9+	9+	0	0	9	49°51′	124°32′
Sechelt	1978	+	31	27	2.6		11	49°30′	123°47′
Sechelt	1979	+	39	35	2.8		11		
Sechelt	1980	+	44	35	3		28		
Sechelt, Acorn Street	1993	+	11	11	2.0	0.6	11	49°28′	123°48′
Sechelt, Acorn Street	1994	+	1						
Squamish, Alice Lake	1993	+	1					49°47′	123°07′
Sutton Islets	1993	+	10+					49°46′	123°58′
Sutton Islets	1994	+							
West Coast and Queen Charlotte Islands									
Kunghit Island	1946	+						52°00′	131°00′
Massett	1920	+						54°00′	132°07′
Massett Inlet	1927	+	large					53°43′	132°20′

Source: Butler 1989; Butler et al. 1995; Campbell et al. 1990; Forbes et al. 1985; Gebauer, unpubl. data; Mark 1976.

APPENDIX 2

Length-mass regression equations of fish caught in beach seines on Roberts Bank, Fraser River delta, BC

Species	Regression equation	*n*	r^2
Three-spined stickleback	lnMass = 0.547 Length – 2.769	46	0.853
Tube-snout	lnMass = 0.332 Length – 3.144	61	0.948
Staghorn sculpin	lnMass = 3.014 lnLength – 4.484	51	0.994
Bay pipefish	lnMass = 3.188 lnLength – 8.354	124	0.738
Shiner seaperch	lnMass = 0.334 Length – 5.49	63	0.925
Gunnels	lnMass = 0.344 Length – 2.607	21	0.873

Note: Scientific names of fish appear in Appendix 4.

APPENDIX 3

Effect of increased disturbance on heron populations

Chris Hitchcock created a mathematical model of the impact of various degrees of disturbance on heron reproductive success using data from many heron studies. The model estimates the number of breeding females alive in the Strait of Georgia following repeated disturbance. In the model, she assumed that:

(1) all herons first bred when they were two years old and all adults bred every year
(2) the annual survival rate of herons from fledging to their second birthday was 45 per cent; survival from their second and third birthdays was 63.7 per cent; and survival from their third birthday and older is 78.1 per cent
(3) that 70 per cent of the young that survived to breed returned to the same size colony (small or large) in which they were born.

Each of these assumptions will introduce error into the model. For example, most herons first begin to breed after their second birthday (Butler 1992a), but we do not know how many begin at a later age (Fernandez-Cruz and Campos 1993). Survival estimates were modified from Henny's (1972) analysis from band-recovery data. This method is based on several assumptions that cannot always be justified (Lakahani and Newton 1983). However, I found similar estimates of local survival using census data of age classes in the Fraser River delta (Butler 1995). Henny (1972) estimated first-year survival from the time of banding until the first birthday to be 31 per cent. Since we use data on the number of chicks fledged, rather than on the number of chicks alive a few weeks earlier, when banding occurs, Henny's estimate will be low. Therefore, we assumed that first-year survival was about 45 per cent. We used Henny's (1972) estimates of survival for all other age classes. The proportion of herons that moved between colonies is unknown, although exchanges of marked individuals occur in this population (Simpson 1984; Simpson et al. 1987). We chose a value of 30 per cent, which is slightly higher than my estimate of 24 per cent for the Point Roberts colony using a different method. Estimates for other colonies are unknown. Since the model was sensitive to migration rates below about 20 per cent, and our estimate for one colony was 24 per cent, we chose 30 per cent as an overall estimate for all colonies.

We started with 530 juveniles, 332 yearlings, and 1,100 adult females in large colonies, and 348 juveniles, 230 yearlings, and 970 adult females in small colonies, for a total of 3,510 female herons. The number of breeding adults corresponds with the estimated number of adults counted in colonies in the early 1990s (see Table 2). To estimate the number of fledglings raised in small and large colonies, a random sample was drawn from four years of data of the mean number of fledglings raised per nesting family in large and small colonies between 1988 and 1991 (Butler et al. 1995). We assumed that increased disturbance would result in fewer chicks produced, both because some birds would not breed at all and because some of the remaining birds would fledge fewer young, as I observed in the Sidney colony in 1987 and 1988. The net decline, if nothing else changed, was about 0.5 per cent per year. With a 25 per cent and 50 per cent increase in disturbances at small colonies, the respective net population declines were about 1.6 per cent and 2.4 per cent per year.

APPENDIX 4

List of common and scientific names used in text

Plants	
Eelgrass	*Zostera marina* and *Z. japonica*
Glasswort	*Salicornia sp.*
Spikegrass	*Distichlis spicata*
Pink sand-verbena	*Abronia latifolia*
Douglas-fir	*Pseudotsuga menziesii*
Red alder	*Alnus rubra*
Sitka spruce	*Picea sitchensis*
Fish	
Pacific herring	*Clupea harengus pacifica*
Shiner perch	*Cymatogaster aggregata*
Three-spined stickleback	*Gasterosteus aculeatus*
Bay pipefish	*Signathus griseolineatus*
Staghorn sculpin	*Leptocottus armatus*
Saddleback gunnel	*Pholis ornata*
Penpoint gunnel	*Apodichthys flavidus*
Starry flounder	*Platychthys flavidus*
Tube-snout	*Aulorhynchus flavidus*
Birds	
Cocoi heron	*Ardea cocoi*
Grey heron	*Ardea cinerea*
Great egret	*Ardea alba*
Reddish egret	*Egretta rufescens*
Tricolored heron	*Egretta tricolor*
Little Blue heron	*Egretta caerulea*
Snowy egret	*Egretta thula*
Double-crested cormorant	*Phalacrocorax auritus*
Bald eagle	*Haliaetus leucocephalus*
Turkey vulture	*Cathartes aura*
American wigeon	*Anas americana*
Mallard	*Anas platyrhynchos*
Northern pintail	*Anas acuta*
Western sandpiper	*Calidris mauri*
Dunlin	*Calidris alpina pacifica*
Northwestern crow	*Corvus caurinus*
Common raven	*Corvus corvax*
Mammals	
Townsend's vole	*Microtus townsendii*
Fallow deer	*Dama dama*

REFERENCES

American Ornithologists' Union. 1983. *Checklist of North American Birds.* Kansas: Allen Press

American Ornithologists' Union. 1996. 'Fortieth Supplement to the American Ornithologists' Union Check-List of North American Birds.' *Auk* 112:819-30

Anderson, J.M. 1978. 'Protection and Management of Wading Birds.' In *Wading Birds*, ed. A. Sprunt IV, J.C. Ogden, and S. Winkler (99-104). National Audubon Society Research Report No. 7. New York: National Audubon Society

Baldwin, J.R., and J.R. Lovvorn. 1994. 'Habitats and Tidal Accessibility of the Marine Foods of Dabbling Ducks and Brant in Boundary Bay, British Columbia.' *Marine Biology* 120:627-38

Bayer, R.D. 1978. 'Aspects of an Oregon Estuarine Great Blue Heron Population.' In *Wading Birds*, ed. A. Sprunt IV, J.C. Ogden, and S. Winkler (213-18). National Audubon Society Research Report No. 7. New York: National Audubon Society

——. 1981. 'Regional Variation of Great Blue Heron Longevity.' *Journal of Field Ornithology* 52:210-13

Bennett, D.C. 1993. 'Growth and Energy Requirements of Captive Great Blue Herons *(Ardea herodias)*.' MSc thesis, University of British Columbia, Vancouver

Bennett, D.C., P.E. Whitehead, and L.E. Hart. 1995. 'Growth and Energy Requirements of Hand-Reared Great Blue Heron (*Ardea herodias*) Chicks.' *Auk* 112:201-9

Bent, A.C. 1926. 'Life Histories of North American Birds.' US National Museum Bulletin 135. Repr. 1963. New York: Dover

Birkhead, T.R. 1991. *The Magpies.* London: T. and A.D. Poyser

Blaker, D. 1969. 'The Behaviour of the Cattle Egret *Ardeola ibis*.' *Ostrich* 40:75-129

Blus, L.J., C.J. Henny, and T.E. Kaiser. 1980. 'Pollution Ecology of Breeding Great Blue Herons in the Columbia Basin, Oregon and Washington.' *Murrelet* 61:63-71

Boonstra, R. 1977. 'Effects of Conspecifics on Survival During Population Declines in *Microtus townsendii*.' *Journal of Animal Ecology* 46:835-51

Boulinier, T., and R.W. Butler. 1997. 'Reproductive Success as a Factor Affecting Settlement Decisions in a Great Blue Heron Metapopulation.' *Oikos* (forthcoming)

Brandman, M. 1976. 'A Quantitative Analysis of the Annual Cycle of Behavior in the Great Blue Heron (*Ardea herodias*).' PhD diss., University of California, Los Angeles

British Columbia Round Table on the Environment and the Economy. 1993. *Georgia Basin Initiative: Creating a Sustainable Future.* Victoria: Ministry of the Environment

Butler, R.W. 1989. 'Breeding Ecology and Population Trends of the Great Blue Heron (*Ardea herodias fannini*) in the Strait of Georgia, British Columbia.' In *The Status and Ecology of Marine and Shoreline Birds in the Strait of Georgia, British Columbia*, ed. K. Vermeer and R.W. Butler (112-17). Ottawa: Canadian Wildlife Service

——. 1991. 'Habitat Selection and Time of Breeding in the Great Blue Heron (*Ardea herodias*).' PhD diss., University of British Columbia, Vancouver

——. 1992a. 'Great Blue Heron.' In *The Birds of North America*, vol. 25, ed. A. Poole, P. Stettenheim, and F. Gill (1-20). Philadelphia: Academy of Natural Science; Washington, DC: American Ornithologists' Union

——. 1992b. *Abundance, Distribution and Conservation of Birds in the Vicinity of Boundary Bay, British Columbia.* Canadian Wildlife Service Technical Report No. 155. Delta, BC: Canadian Wildlife Service

——. 1993. 'Time of Breeding in Relation to Food Availability of Female Great Blue Herons

(*Ardea herodias*).' *Auk* 110:693-701

——. 1994a. 'Population Regulation in Wading Ciconiiform Birds.' *Colonial Waterbirds* 17:189-99

——. 1994b. 'Distribution and Abundance of Western Sandpipers, Dunlins and Black-Bellied Plovers in the Fraser River Estuary.' In *The Abundance and Distribution of Waterbirds in Estuaries in the Strait of Georgia*, ed. R.W. Butler and K. Vermeer (24-36). Canadian Wildlife Service Occasional Paper No. 83. Ottawa: Canadian Wildlife Service

——. 1995. *The Patient Predator: Population and Foraging Ecology of the Great Blue Heron* (Ardea herodias) *in British Columbia.* Canadian Wildlife Service Occasional Paper No. 83. Ottawa: Canadian Wildlife Service

Butler, R.W., A.M. Breault, and T.M. Sullivan. 1990. 'Measuring Animals through a Telescope.' *Journal of Field Ornithology* 61:111-14

Butler, R.W., and R.W. Campbell. 1987. *The Birds of the Fraser River Delta: Populations, Ecology and International Significance.* Canadian Wildlife Service Occasional Paper No. 65. Ottawa: Canadian Wildlife Service

Butler, R.W., and R.J. Cannings. 1989. *Distribution and Abundance of Birds in the Intertidal Portion of the Fraser River Delta, British Columbia.* Canadian Wildlife Service Technical Report No. 93. Delta, BC: Canadian Wildlife Service

Butler, R.W., F.S. Delgado, H. de la Cueva, V. Pulido, and B.K. Sandercock. 1996. 'Migration Routes of the Western Sandpiper.' *Wilson Bulletin* 108:662-72

Butler, R.W., G.W. Kaiser, and G.E.J. Smith. 1987. 'Migration Chronology, Length of Stay, Sex Ratio, and Weight of Western Sandpipers (*Calidris mauri*) on the South Coast of British Columbia.' *Journal of Field Ornithology* 58:103-11

Butler, R.W., and J.W. Kirbyson. 1979. 'Oyster Predation by the Black Oystercatcher in British Columbia.' *Condor* 81:433-5

Butler, R.W., N.A.M. Verbeek, and H. Richardson. 1984. 'The Breeding Biology of the Northwestern Crow.' *Wilson Bulletin* 96:408-18

Butler, R.W., and K. Vermeer. 1989. 'Overview and Recommendations: Important Bird Habitats and the Need for Their Protection.' In *The Status and Ecology of Marine and Shoreline Birds in the Strait of Georgia, British Columbia*, ed. K. Vermeer and R.W. Butler (185-6). Ottawa: Canadian Wildlife Service

——. 1994. *The Abundance and Distribution of Waterbirds in Estuaries in the Strait of Georgia*, ed. R.W. Butler and K. Vermeer. Canadian Wildlife Service Occasional Paper No. 83. Ottawa: Canadian Wildlife Service

Butler, R.W., P.E. Whitehead, A.M. Breault, and I.E. Moul. 1995. 'Colony Effects on Fledging Success of Great Blue Herons (*Ardea herodias*) in British Columbia.' *Colonial Waterbirds* 18:159-65

Campbell, R.W., N.K. Dawe, I. McTaggart-Cowan, J.M. Cooper, G.W. Kaiser, and M.C.E. McNall. 1990. *The Birds of British Columbia.* Vol. 1. Victoria: Royal British Columbia Museum

Cannings, R.J., and S.G. Cannings. 1996. *British Columbia: A Natural History.* Vancouver: Greystone Press

Caraco, T.S. 1981. 'Risk-Sensitivity and Foraging Groups.' *Ecology* 62:527-31

Caraco, T.S., S. Martindale, and T.S. Whitham. 1980. 'An Empirical Demonstration of Risk-Sensitive Foraging Preferences.' *Animal Behavior* 28:820-30

Carlson, B.A., and E.B. McLean. 1996. 'Buffer Zone and Disturbance Types as Predictors of Fledging Success in Great Blue Herons, *Ardea herodias*.' *Colonial Waterbirds* 19:124-7

Chitty, D. 1967. 'The Natural Selection of Self-regulatory Behaviour in Animal Populations.'

Proceedings of the Ecological Society of Australia 2:51-78

Clobert, J., and J.-D. Lebreton. 1991. 'Estimation of Demographic Parameters in Bird Populations.' In *Bird Population Studies: Relevance to Conservation and Management*, ed. C.M. Perrins, J.-D. Lebreton, and G.J.M. Hirons (75-104). Oxford: Oxford University Press

Collazo, J.A. 1981. 'Some Aspects of the Breeding Ecology of the Great Blue Heron at Heyburn State Park.' *Northwest Science* 55:293-7

Cottrille, W.P., and B.D. Cottrille. 1958. *Great Blue Heron: Behavior at the Nest.* University of Michigan Miscellaneous Publications 102. Ann Arbor: University of Michigan

Darwin, C. 1871. *The Descent of Man and Selection in Relation to Sex.* London: Murray

Davis, W. 1992. *Shadows in the Sun: Essays on the Spirit of Place.* Edmonton: Lone Pine Press

DesGranges, J.-L., P. Laporte, and G. Chapdelaine. 1979a. *First Tour of Inspection of Quebec Heronries, 1977.* Canadian Wildlife Service Program Notes No. 93. Ottawa: Canadian Wildlife Service

—. 1979b. *Second Tour of Inspection of Quebec Heronries, 1978.* Canadian Wildlife Service Program Notes No. 105. Ottawa: Canadian Wildlife Service

Dolesh, R.J. 1984. 'Lord of the Shallows, the Great Blue Heron.' *National Geographic* 165:540-54

Dowd, E.M., and L.D. Flake. 1985. 'Foraging Habitat and Movements of Nesting Great Blue Herons in a Prairie River Ecosystem, South Dakota.' *Journal of Field Ornithology* 56:379-87

Downes, C., and B. Collins. 1996. *The Canadian Breeding Bird Survey, 1966-1994.* Canadian Wildlife Service Program Notes No. 210. Ottawa: Canadian Wildlife Service

Drent, R.H., and S. Daan. 1980. 'The Prudent Parent: Energetic Adjustments in Avian Breeding.' *Ardea* 68:225-52

Dunn, E., D.J.T. Hussell, and J. Siderius. 1985. 'Status of the Great Blue Heron, *Ardea herodias*, in Ontario.' *Canadian Field Naturalist* 99:62-70

Eissinger, A. 1996. 'Great Blue Herons of the Salish Sea: A Model Plan for the Conservation and Stewardship of Coastal Heron Colonies.' Report prepared for Trillium Corporation, ARCO Products, and Washington Department Fish and Wildlife, Olympia, WA

Elliott, J.E., R.W. Butler, R.J. Nostrom, and P.E. Whitehead. 1989. 'Environmental Contaminants and Reproductive Success of Great Blue Heron (*Ardea herodias*) in British Columbia, 1986-87.' *Environmental Pollution* 59:91-114

Elliott, J.E., P.E. Whitehead, P.A. Martin, G.D. Bellward, and R.J. Norstrom. 1996. 'Persistent Pulp Mill Pollutants in Wildlife.' In *Environmental Fate and Effects of Pulp and Paper Mill Effluents*, ed. M.R. Servos, K.R. Munkittrick, J.H. Carey, and G.J. Van der Kraak (297-314). Delray Beach, FL: St. Lucie Press

English, S.M. 1978. 'Distribution and Ecology of Great Blue Heron Colonies on the Willamette River, Oregon.' In *Wading Birds*, ed. A. Sprunt IV, J.C. Ogden, and S. Winkler (235-46). National Audubon Society Research Report No. 7. New York: National Audubon Society

Erwin, R.M., J.G. Haig, D.B. Stotts, and J.S. Hatfield. 1996. 'Reproductive Success, Growth and Survival of Black-Crowned Night-Heron (*Nycticorax nycticorax*) and Snowy Egret (*Egretta thula*) Chicks in Coastal Virginia.' *Auk* 113:119-30

Fernandez-Cruz, M., and F. Campos. 1993. 'The Breeding of Grey Herons (*Ardea cinera*) in Western Spain: The Influence of Age.' *Colonial Waterbirds* 16:53-8

Forbes, L.S. 1989. 'Coloniality in Herons: Lack's Predation Hypothesis Reconsidered.' *Colonial Waterbirds* 12:24-9

Forbes, L.S., and K. Simpson. 1985. 'Behavioural Studies of Great Blue Herons at Pender Harbour and Sechelt, British Columbia in 1980.' Report prepared for Canadian Wildlife Service, Delta, BC

Forbes, L.S., K. Simpson, J.P. Kelsall, and D.R. Flook. 1985a. 'Great Blue Heron Colonies in British Columbia.' Report prepared for Canadian Wildlife Service, Delta, BC

—. 1985b. 'Reproductive Success of Great Blue Herons in British Columbia.' *Canadian Journal of Zoology* 63:1110-13

Fraser Basin Management Program. 1996. *Board Report Card 1996*. Vancouver: Fraser Basin Management Program

Gaston, A.J. 1992. *The Ancient Murrelet: A Natural History in the Queen Charlotte Islands.* London: T. and A.D. Poyser

Gebauer, M.B. 1993. *Status and Productivity of Great Blue Heron* (Ardea herodias) *Colonies in the Lower Fraser River Valley, 1992.* Surrey, BC: BC Ministry of Environment, Lands, and Parks

Gibbs, J.P. 1991. 'Spatial Relationships between Nesting Colonies and Foraging Areas of Great Blue Herons.' *Auk* 108:764-70

Gibbs, J.P., S. Woodward, M.L. Hunter, and A.E. Hutchinson. 1987. 'Determinants of Great Blue Heron Colony Distribution in Coastal Maine.' *Auk* 104:38-47

Goodman, D. 1974. 'Natural Selection and a Cost Ceiling on Reproductive Effort.' *American Naturalist* 108:247-68

Gordon, D.K., and C.D. Levings. 1984. *Seasonal Changes of Inshore Fish Populations on Sturgeon and Roberts Banks, Fraser River Estuary, British Columbia.* Canadian Technical Report Fisheries and Aquatic Sciences 1240. West Vancouver, BC: Department of Fisheries and Oceans

Gutsell, R. 1995. 'Age-Related Foraging Behaviour and Habitat Use in Great Blue Herons.' MSc thesis, Simon Fraser University, Burnaby, BC

Hancock, J., and H. Elliott. 1978. *The Herons of the World.* London: London Editions

Hancock, J., and J. Kushlan. 1984. *The Herons Handbook.* New York: Harper and Row

Harcombe, A., W. Harper, S. Cannings, D. Fraser, and W.T. Munro. 1994. 'Terms of Endangerment.' In *Biodiversity in British Columbia*, ed. L.E. Harding and E. McCullum (11-28). Ottawa: Supply and Services

Harding, L.E., and E. McCullum. 1994. *Biodiversity in British Columbia.* Ottawa: Supply and Services

Harfenist, A., P.E. Whitehead, W.J. Cretney, and J.E. Elliott. 1995. *Food Chain Sources of Polychlorinated Dioxins and Furans to Great Blue Herons* (Ardea herodias) *Foraging in the Fraser River Estuary, British Columbia.* Canadian Wildlife Service Technical Report No. 169. Delta, BC: Canadian Wildlife Service

Harwood, M. 1977. *Moments of Discovery: Adventures with North American Birds.* New York: E.P. Dutton

Henny, C.J. 1972. *An Analysis of the Population Dynamics of Selected Avian Species with Special Reference to Changes during the Modern Pesticide Era.* US Fish and Wildlife Service, Wildlife Research Report 1. Washington, DC: US Fish and Wildlife Service

Henny, C.J., and M.R. Bethers. 1971. 'Population Ecology of the Great Blue Heron, with Special Reference to Western Oregon.' *Canadian Field Naturalist* 85:205-9

Hobson, K.A., and J.C. Driver. 1989. 'Archaeological Evidence for Use of the Strait of Georgia by Marine Birds.' In *The Status and Ecology of Marine and Shoreline Birds in the Strait of Georgia, British Columbia*, ed. K. Vermeer and R.W. Butler (168-73). Canadian Wildlife Service Special Publication. Ottawa: Canadian Wildlife Service

Hughes, G.W. 1985. 'The Comparative Ecology and Evidence for Resource Partitioning in Two Pholidid Fishes *(Pisces: Pholididae)* from Southern British Columbia Eelgrass Beds.' *Canadian Journal of Zoology* 63:76-85

Hutchison, B. 1950. *The Fraser.* Toronto: Clarke, Irwin

Jones, I.L., and F.M. Hunter. 1993. 'Mutual Sexual Selection in a Monogamous Seabird.' *Nature* 362:238-9
Julin, K.R. 1986. 'Decline of Second-Growth Douglas Fir in Relation to Great Blue Heron Nesting.' *Northwest Science* 60:201-5
Kelsall, J.P., and K. Simpson. 1979. 'A Three Year Study of the Great Blue Heron in Southwestern British Columbia.' *Proceedings of the Colonial Waterbird Group* 3:69-74
Kendeigh, S.C. 1963. 'Regulation of Nesting Time and the Distribution in the House Wren.' *Wilson Bulletin* 75:418-27
Ketchen, K.S., N. Bourne, and T.H. Butler. 1983. 'History and Present Status of Fisheries for Marine Fishes and Invertebrates in the Strait of Georgia, British Columbia.' *Canadian Journal of Fisheries and Aquatic Sciences* 40:1095-119
Koonz, W.H., and P.W. Rakowski. 1985. 'Status of Colonial Waterbirds Nesting in Southern Manitoba.' *Canadian Field Naturalist* 99:19-29
Krebs, C.J. 1979. 'Dispersal, Spacing Behaviour, and Genetics in Relation to Population Fluctuations in the Vole, *Microtus townsendii*.' *Fortschritte der Zoologie* 25:61-77
Krebs, E.A. 1991. 'Reproduction in the Cattle Egret (*Bubulcus ibis*): The Function of Breeding Plumes.' MSc thesis, McGill University, Montreal
Krebs, J.R. 1974. 'Colonial Nesting and Social Feeding as Strategies for Exploiting Food Resources in the Great Blue Heron (*Ardea herodias*).' *Behaviour* 51:99-131
Kushlan, J.A. 1976. 'Site Selection for Nesting Colonies by the American White Ibis *Eudocimus albus* in Florida.' *Ibis* 118:590-3
Lack, D. 1954. *The Natural Regulation of Animal Numbers.* Oxford: Clarendon Press
Lakahani, K.H., and I. Newton. 1983. 'Estimating Age-Specific Bird Survival Rates from Ring Recoveries – Can It Be Done?' *Journal of Animal Ecology* 52:83-91
Lancaster, D.A. 1970. 'Breeding Behaviour of Cattle Egret in Colombia.' *Living Bird* 9:167-94
Leach, B.A. 1972. 'The Waterfowl of the Fraser Delta, British Columbia.' *Wildfowl* 23:45-55
Lekuona, J.-M., and F. Campos. 1995. 'Le succès de reproduction du heron cendre *Ardea cinerea* dans le basin d'Arcachon.' *Aluda* 63:179-83
Lovvorn, J.R., and J.R. Baldwin. 1996. 'Intertidal and Farmland Habitats of Ducks in the Puget Sound Region: A Landscape Perspective.' *Biological Conservation* 77:97-114
Mangel, M., et al. 1996. 'Principles for the Conservation of Wild Living Resources.' *Ecological Applications* 6:338-62
Marion, L. 1989. 'Territorial Feeding and Colonial Breeding Are Not Mutually Exclusive: The Case of the Grey Heron (*Ardea cinerea*).' *Journal of Animal Ecology* 58:693-710
Mark, D.M. 1976. 'An Inventory of Great Blue Heron (*Ardea herodias*) Nesting Colonies in British Columbia.' *Northwest Science* 50:32-41
McAloney, K. 1973. 'The Breeding Biology of the Great Blue Heron on Tobacco Island, Nova Scotia.' *Canadian Field Naturalist* 87:137-40
Meyerriecks, A.J. 1960. *Comparative Breeding Behavior of Four Species of North American Herons.* Publications of the Nuttall Ornithological Club No. 2. Cambridge, MA: Nuttall Ornithological Club
Milstein, P.L., I. Prestt, and A.A. Bell. 1970. 'The Breeding Cycle of the Grey Heron.' *Ardea* 58:171-257
Mock, D.W. 1976. 'Pair Formation Displays of the Great Blue Heron.' *Wilson Bulletin* 88:185-230
—. 1978. 'Pair-Formation Displays of the Great Blue Heron.' *Condor* 80:159-72
—. 1979. 'Display Repertoire Shifts and "Extra-Marital" Courtship in Herons.' *Behaviour* 69:57-71

——. 1985. 'Siblicidal Brood Reduction: The Prey Size Hypothesis.' *American Naturalist* 125:327-43

Mock, D.W., T.C. Lamey, and D.B.A. Thompson. 1988. 'Falsifiability and the Information Centre Hypothesis.' *Ornis Scandinavica* 19:231-48

Moore, K. 1990. *Urbanization in the Lower Fraser Valley, 1980-1987.* Canadian Wildlife Service Technical Report No. 120. Delta, BC: Canadian Wildlife Service

Moul, I.E. 1990. 'Environmental Contaminants, Disturbance and Breeding Failure at a Great Blue Heron Colony on Vancouver Island.' MSc thesis, University of British Columbia, Vancouver

Nillsson, J.-A. 1995. 'Parent-Offspring Interaction over Brood Size: Cooperation or Conflict?' *Journal of Avian Biology* 26:255-9

Norman, D.M. 1995. 'The Status of Great Blue Herons in Puget Sound: Population Dynamics and Recruitment Hypotheses.' In *Puget Sound Resources '95 Proceedings*, ed. Olympia, WA: Puget Sound Water Quality Authority

Norman, D.M., A.M. Breault, and I.E. Moul. 1989. 'Bald Eagle Incursions and Predation at Great Blue Heron Colonies.' *Colonial Waterbirds* 12:143-230

North, M.E.A., and J.M. Teversham. 1984. 'The Vegetation of the Lower Fraser, Serpentine and Nicomekl Rivers, 1859 to 1890.' *Syesis* 17:47-66

Payne, R.B. 1979. '*Ardeidae*.' In *Checklist of Birds of the World*, ed. E. Mayr and G.W. Cottrell (193-244). Cambridge, MA: Museum of Comparative Zoology

Pearse, T. 1968. *Birds of the Early Explorers in the Northern Pacific.* Comox, BC: n.p.

Perrins, C.M., and T.R. Birkhead. 1983. *Avian Ecology.* London: Blackie

Powell, G.V.N. 1983. 'Food Availability and Reproduction by Great White Herons *Ardea herodias*: A Food Addition Study.' *Colonial Waterbirds* 6:138-47

Powell, G.V.N., and A.H. Powell. 1986. 'Reproduction by Great White Herons *Ardea herodias* in Florida Bay as an Indicator of Habitat Quality.' *Biological Conservation* 36:101-13

Pratt, H.M. 1970. 'Breeding Biology of Great Blue Herons and Common Egrets in Central California.' *Condor* 72:407-16

Pratt, H.M., and D.W. Winkler. 1985. 'Clutch Size, Timing of Laying and Reproductive Success in a Colony of Great Blue Herons and Great Egrets.' *Auk* 102:49-63

Quinney, T.E., and P.C. Smith. 1979. *Reproductive Success, Growth of Nestlings and Foraging Behaviour of the Great Blue Heron* (Ardea herodias herodias L.). Canadian Wildlife Service Contract Report No. KL229-5-7077. Ottawa: Canadian Wildlife Service

——. 1980. 'Comparative Foraging Behaviour and Efficiency of Adult and Juvenile Great Blue Herons.' *Canadian Journal of Zoology* 58:1168-73

Richner, H. 1986. 'Winter Feeding Strategies of Individually Marked Herons.' *Animal Behavior* 34:881-6

Sealy, S.G. 1973. 'Interspecific Feeding Assemblages of Marine Birds off British Columbia.' *Auk* 90:796-802

Shugart, G.W., and S. Rohwer. 1996. 'Serial Descendent Primary Molt or *Staffelmauser* in Black-Crowned Night-Herons.' *Condor* 98:222-33

Siegfried, W.R. 1971. 'Plumage and Moult of the Cattle Egret.' *Ostrich Supplement* 9:153-64

Simenstad, C. 1983. *The Ecology of Estuarine Channels of the Pacific Northwest: A Community Profile.* Washington, DC: US Fish and Wildlife Service, US Department of the Interior

Simpson, K. 1984. 'Factors Affecting Reproduction in Great Blue Herons (*Ardea herodias*).' MSc thesis, University of British Columbia, Vancouver

Simpson, K., J.N.M. Smith, and J.P. Kelsall. 1987. 'Correlates and Consequences of Coloniality in Great Blue Herons.' *Canadian Journal of Zoology* 65:572-7

Snow, D.W. 1955. 'The Abnormal Breeding of Birds in the Winter 1953-54.' *British Birds* 48:121-6

Stenning, M.J. 1996. 'Hatching Asynchrony, Brood Reduction and Other Rapidly Reproducing Hypotheses.' *Trends in Ecology and Evolution.* 6:243-6

Stephens, D.W., and J.R. Krebs. 1986. *Foraging Theory.* Princeton: Princeton University Press

Taitt, M.J. 1984. 'Experimental Analysis of Spacing Behaviour in the Vole, *Microtus townsendii.*' In *Behavioural Ecology*, ed. R.M. Sibly and R.H. Smith (313-17). London: Blackwell

Taitt, M.J., J.H.W. Gipps, C.J. Krebs, and Z. Dundjerski. 1981. 'The Effect of Extra Food and Cover on Declining Populations of *Microtus townsendii.*' *Canadian Journal of Zoology* 59:1593-9

Taitt, M.J., and C.J. Krebs. 1983. 'Predation, Cover and Food Manipulation during a Spring Decline of *Microtus townsendii.*' *Journal of Animal Ecology* 51:413-28

Triton Consultants Ltd. 1995. *A Review of Recent Physical and Biological Development of the Southern Roberts Bank Seagrass System, 1950-1994.* Vol. 1. Vancouver: Roberts Bank Environmental Review Committee, Environment Canada

van Vessem, J., and D. Draulans. 1986. 'The Adaptive Significance of Colonial Breeding in the Grey Heron *Ardea cinerea*: Inter- and Intra-Colony Variability in Breeding Success.' *Ornis Scandinavica* 17:356-62

Verbeek, N.A.M., and R.W. Butler. 1989. 'Feeding Ecology of Shoreline Birds in the Strait of Georgia, British Columbia.' In *The Ecology and Status of Marine and Shoreline Birds in the Strait of Georgia, British Columbia*, ed. K. Vermeer and R.W. Butler (74-81). Canadian Wildlife Service Special Publication. Ottawa: Canadian Wildlife Service

Vermeer, K. 1969. 'Great Blue Heron Colonies in Alberta.' *Canadian Field Naturalist* 83:237-42

Vermeer, K., M. Bentley, and K.H. Morgan. 1994. 'Comparison of the Waterbird Populations of the Chemainus, Cowichan and Nanaimo River Estuaries.' In *The Abundance and Distribution of Estuarine Birds in the Strait of Georgia, British Columbia*, ed. R.W. Butler and K. Vermeer (57-62). Canadian Wildlife Service Occasional Paper No. 83. Ottawa: Canadian Wildlife Service

Vermeer, K., and R.W. Butler. 1994. 'The International Significance and the Need for Environmental Knowledge of the Estuaries in the Strait of Georgia.' In *The Abundance and Distribution of Waterbirds in Estuaries in the Strait of Georgia*, ed. R.W. Butler and K. Vermeer (75-6). Canadian Wildlife Service Occasional Paper No. 83. Ottawa: Canadian Wildlife Service

Vermeer, K., R.W. Butler, and K.H. Morgan. 1994. 'Comparison of Seasonal Shorebird and Waterbird Densities within the Fraser River Delta Intertidal Regions.' In *The Abundance and Distribution of Waterbirds in Estuaries in the Strait of Georgia*, ed. R.W. Butler and K. Vermeer (6-17). Canadian Wildlife Service Occasional Paper No. 83. Ottawa: Canadian Wildlife Service

Vermeer, K., K.H. Morgan, R.W. Butler, and G.E.J. Smith. 1989. 'Population, Nesting Habitat and Food of Bald Eagles in the Gulf Islands.' In *The Status and Ecology of Marine and Shoreline Birds in the Strait of Georgia, British Columbia*, ed. K. Vermeer and R.W. Butler (123-31). Canadian Wildlife Service Special Publication. Ottawa: Canadian Wildlife Service

Vermeer, K., I. Robertson, R.W. Campbell, G. Kaiser, and M. Lemon. 1983. *Distribution and Densities of Marine Birds on the Canadian West Coast.* Delta, BC: Environment Canada

Vos, D.K., D.A. Ryder, and W.D. Graul. 1985. 'Response of Breeding Great Blue Herons to Human Disturbance in Northcentral Colorado.' *Colonial Waterbirds* 8:13-22

Werschkul, D.F., E. McMahon, and M. Leitschuh. 1976. 'Some Effects of Human Activities on the Great Blue Heron in Oregon.' *Wilson Bulletin* 88:660-2

——. 1977. 'Observations on the Reproductive Ecology of the Great Blue Heron in Western Oregon.' *Murrelet* 58:7-12

White, H., and J. Spilsbury. 1987. *Spilsbury's Coast.* Madeira Park, BC: Harbour Publishing

Whitehead, P.E. 1989. 'Toxic Chemicals in Great Blue Herons' (*Ardea herodias*) Eggs in the Strait of Georgia.' In *The Status and Ecology of Marine and Shoreline Birds in the Strait of Georgia, British Columbia,* ed. K. Vermeer and R.W. Butler (177-83). Canadian Wildlife Service Special Publication. Ottawa: Canadian Wildlife Service

Wiebe, J.P. 1968. 'The Reproductive Cycle of the Viviparous Sea Perch, *Cymatogaster aggregata.*' *Canadian Journal of Zoology* 46:1221-34

Wiese, J.H. 1976. 'Courtship and Pair Formation in the Great Egret.' *Auk* 93:709-24

Williams, G.C. 1966. *Adaptation and Natural Selection.* Princeton: Princeton University Press

Woolfenden, G.E., S.C. White, R.L. Mumme, and W.B. Robertson Jr. 1976. 'Aggression among Starving Cattle Egrets.' *Bird Banding* 47:48-53

Zammuto, R.M. 1986. 'Life Histories of Birds: Clutch Size, Longevity, and Body Mass among North American Game Birds.' *Canadian Journal of Zoology* 64:2739-49

INDEX

Note: all entries refer to the great blue heron *(Ardea herodias fannini)* unless otherwise specified. A 't' after a page number indicates a table; 'f' indicates a figure; 'p' indicates a photograph.

Set in Minion with Centaur display
Printed and bound in Canada by Friesens
Copy-editor: Dallas Harrison
Proofreader: Maureen Nicholson
Cartographer: Eric Leinberger
Indexer: Patricia Buchanan
Designer: George Vaitkunas